D1013062

BLOCKCHAIN:
THE NEXT
EVERYTHING

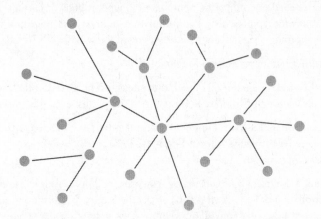

Stephen P. Williams

SCRIBNER

New York London Toronto Sydney New Delhi

Scribner
An Imprint of Simon & Schuster, Inc.
1230 Avenue of the Americas
New York, NY 10020

First Scribner hardcover edition March 2019

SCRIBNER and design are registered trademarks of The Gale Group, Inc., used under license by Simon & Schuster, Inc., the publisher of this work.

For information about special discounts for bulk purchases, please contact Simon & Schuster Special Sales at 1-866-506-1949 or business@simonandschuster.com.

The Simon & Schuster Speakers Bureau can bring authors to your live event. For more information or to book an event, contact the Simon & Schuster Speakers Bureau at 1-866-248-3049 or visit our website at www.simonspeakers.com.

Interior design by Jason Snyder

Manufactured in the United States of America

10 9 8 7 6 5 4 3 2 1

Library of Congress Cataloging-in-Publication Data is available.

ISBN 978-1-9821-1682-8
ISBN 978-1-9821-1684-2 (ebook)

For the brighter future of
Bolivia, Violet, and Aspen

CONTENTS

There is something new and fundamental happening in the world which could be the start of the next enlightenment period. The core of this is shifting from centralized to decentralized models in all aspects of our lives, both individual and societally.

*—Melanie Swan, a technology theorist in the
Philosophy Department at Purdue University, writing
in "Crypto Enlightenment: A Social Theory of Blockchains,"
a paper published by the Institute for Ethics
and Emerging Technologies*

Introduction

Late summer, 2015, the Hudson River reflecting sunlight up onto the trees along the esplanade. I was reading an article about the perils of doing business in the Democratic Republic of Congo (DRC) — slave labor, corruption, exploitation of endangered habitats to get rare minerals such as coltan, which is essential to making smartphones. And then the word: blockchain. Apparently, the article said, this new technology could serve as a ledger to keep track of coltan that was mined in responsible ways. Blockchain could track sustainably mined coltan step by step on its journey from the forest all the way to the phone. That way, someone in America or Europe who bought a phone could be sure that it hadn't endangered any gorilla habitats, for instance.

That's a good feeling for the person buying the phone, for sure. But what about the miners who dug up that coltan? A little research showed me that blockchain also offered powerful ways of leveling the playing field, so the miners' contributions might be better recognized. An artisanal miner, for instance, who now lost track of his coltan once the local middle agent bought it, would be able to track the movement of his product all along the chain. That transparency might give the miner some ideas on how to improve various stages of that long supply chain stretching from the DRC to the Chinese factories cranking out the phones, and eventually to the Apple Store on West Fourteenth Street, not far from where I sat, watching the river flow.

What kind of economy might develop if the first person on the supply chain had access to the prices at the final stop on the supply chain? How helpful might it be for the corporation using that coltan to receive creative input from the miner, who in many ways is more intimate with the product than anyone? At the very least, an important story could be told that would help the consumer understand the product. And that seemed to be only the beginning of blockchain's potential.

Blockchain—what a weird name for such a cool concept. I soon became obsessed and dove in. This book is where I've come up for air.

———————————

Blockchain technology was originally devised as a platform for bitcoin that would track the spending or sale of each digital coin, transaction by transaction. That function alone would be enough to make the technology a world-changing invention. Yet soon after the introduction of bitcoin, in 2008, technologists began to realize that the underlying blockchain might have even greater value in the long run than the cryptocurrencies it was created for.

Blockchain is a simple technology that, at its most basic, serves as a permanent, unhackable ledger for almost any kind of information you'd like to record. Yet it turns out this simple ledger technology makes an ideal platform for building all sorts of innovative and radically new applications.

For example, register the sale of an acre of land, and there will never be a question of who owns it the next time it is sold. No title company needs to certify the sale, either. This application is already being developed in Honduras, where an estimated 80 percent of private land has an improper title, or no title at all. Blockchain registry of ownership can help prevent land theft, including invasions, and help protect forests from illegal logging and settlements.

Another type of real estate—digital real estate that only exists on the Internet—is being offered in a digital realm called Genesis City. A limited number of "lots" are being sold, with title registered on a blockchain, to people who can develop the digital space however they want. Is this hucksterism on the level of selling Florida swampland to northerners in the 1950s? Or is there value in owning a section of a digital world that is expected to be visited by millions? You could build a store, a game, a magazine, or something else in this space.

The benefits of blockchain extend far beyond simple money-making schemes. In East Africa, a company called Wala uses blockchain to give formerly unbanked people access to banking via their smartphones. This allows them for the first time to participate in the modern economy. And in New York, an artist named Kevin Abosch has recorded blockchain alphanumeric codes with his own blood and offered them for sale. He also sold the cryptographic registration of a photo he took for one million dollars, photo not included.

A growing number of businesses are now being created to operate without the traditional top-down hierarchy, using blockchain to decentralize ownership and control in ways that many people hope herald the coming of a more egalitarian capitalism. Something about blockchain inspires certain people to explore the tech in wildly creative ways.

On the other hand, as I discovered in my journey through the world of blockchain, it's a technology that generates a lot of talk, with very little real understanding of how it works. While I don't believe it's necessary for everyone to have a deep technical understanding of blockchain, a conceptual understanding of how the tech works makes its transformative potential in our society much easier to imagine. I'm sharing what I've learned, here, with the hope of inspiring you to envision how blockchain might begin to transform your own life.

I

EXPERIENCING BLOCKCHAIN

What Is Blockchain?

Blockchain is the tech. Bitcoin is merely the first mainstream manifestation of its potential.

—*Marc Kenigsberg*

Blockchain is software. It's as simple as that.

As software, blockchains are not physical objects. They exist only as strings of code on computers, phones, and other devices. You join a chain by connecting your phone, computer, or other device to it via software and the Internet. By joining, you become part of a system that might include hundreds, thousands, or even millions of connected people and machines, all of you on the same playing field, with no hierarchical control. The ephemeral nature of the chains belies how they can grant individuals unprecedented power in the marketplaces of ideas, governance, and finance. You can't touch a blockchain, but it can touch you.

Lie back on a blanket looking up into the clear night sky. Imagine that each star is connected by a beam of light to the star closest to it. That star is connected to another. And another. So that beams of light connect all the nearby stars. And those stars are connected to their neighbors, and on and on with new stars joining into infinity, so that every star is connected, via other stars, to the star that's closest to you. Your imagination makes the night sky a model of a distributed network. That's where blockchains live. The stars are computers, phones, or other devices, all connected, all sharing the duties of running the computer code that is a blockchain.

What Blockchain Does

Blockchain-based market networks will
replace existing networks. Slowly, then suddenly.
In one thing, then in many things.

—*Naval Ravikant*

Put simply, a blockchain is an online ledger. At heart, it resembles those large green books in which your ancestors recorded the sale of a tractor, or pennies set aside for a rainy day. Somewhat dull, right? Not when you consider that ledgers are the basis of civilization. Without them, we would never have developed markets, or cities. Ledgers:

Record all the data in our economy

Show status, such as citizenship or employment

Confirm membership

Show ownership

Track the value of things

Establish intellectual property ownership

Make art "real" by confirming provenance

Confirm our identity on this earth

In other words, the software is a digital ledger for keeping track of any kind of data. This software is able to track the flow of goods, the movement of money, the provenance of artworks or poems, the treatment of refugees seeking asylum, and the health of the tundra, among other things. Two specific blockchains, called Ethereum and bitcoin, are the most well known and valuable. Yet there are many blockchains, living on millions of computers.

Despite their strong connection in the public mind, blockchain is not bitcoin. Rather, blockchain is a digital ledger that makes bitcoin and other cryptocurrencies possible by recording the existence of each coin, who owns it, when it is sold, and to whom. This is vital, because you can't signal crypto ownership by jangling a bitcoin in your pocket—the coins exist only in the digital realm, as computer code. Without blockchain, bitcoins would have no value.

All blockchains are distinguished by their ability to store information immutably, meaning the information can't be changed or hacked. This feature is built into the design of the technology, for reasons we'll get to shortly, and it's probably the most significant quality of any blockchain. For instance, right now, if you want to wire money to an overseas company to pay for a shipment of widgets, you must depend on your bank to tell its bank (most likely via an intermediary bank of some sort) that the money you are sending is legitimate. It might take several days for your money to actually reach its account. In turn, the overseas company must depend on its bank to certify that legitimacy. Each step of this process has a fee; in effect, you are paying for "trust."

The immutable nature of transactions on a blockchain can eliminate this need for middle agents and payments, because the money has been indisputably recorded as yours to spend. Each time it is transferred, the new owner is recorded. You cannot "double spend" your money, or claim to have money that you don't.

The ledger's certainty leads people to say that blockchain is "trustless," meaning the mechanical nature of the unhackable algorithms makes trust unnecessary.

The basic function of each blockchain is to group digital information into collections, called "blocks," that can't be altered. This information can include anything from a financial transfer to the census count of rubber trees in Amazonia—anything that someone wants to permanently record. When a block is full of data—after perhaps 2,000 entries—one of several complex processes is used to time-stamp it into a permanent record. Each block is linked to the next by a code that references the content of both blocks. That's why we call it a "chain."

If someone tries to change the information in a closed block, the codes linking it to neighboring blocks will no longer be in sync. Virtual alarms will signal a break-in. That's great security. But that's not all.

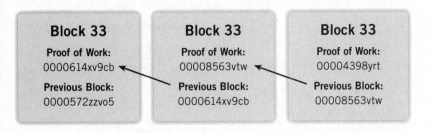

Anyone who joins a blockchain, using his or her device, is considered to be a "node" on the blockchain. There can be thousands, even millions, of nodes on a chain—the more the merrier. The true brilliance of the design is that the blockchain is replicated, and constantly updated, on every one of these nodes. In the remote event that someone were able to change the information in a block, and all the blocks that followed it, that person would also have to do that on every separate node in the chain. Otherwise, virtual alarms again. It would be an impossible task.*

This collection of nodes can be either a decentralized or a distributed network, and both are rich with possibility for business, the arts, social enterprises, and more. Decentralized networks retain some minimal hierarchical structure. In distributed networks, which take the concept of decentralization to another level, every node has a say in how that network is governed—there is no hierarchy, boss, or father figure. Every node is capable of supporting the chain even if some of the nodes around it fail. The distributed network is a radical innovation in how to structure corporations, governments, and other systems that we depend on to maintain society. We do not know where commitment to distributed networks will lead us, though I am optimistic that eventually we will find ourselves in a good place.

---

* Alas, the elusive holy grail of quantum computing poses a threat to this security. Many worry that this type of computing, if realized, would be so powerful that it could potentially crack blockchain technology.

Chances are that you will never have to consciously access a blockchain. Chances are equally good that within five years you will be connecting to blockchains every day, using applications that work seamlessly with the technology, unbeknownst to you, via your device. Blockchain is an underlying technology, a platform that makes other things possible. In a broad sense, it's conceptually similar to how your computer operating system lets you search with Google, or write with Word, or listen to music on Spotify. I'm guessing that most of us don't actually know how these things work. As with other software operating systems, we don't need to understand exactly what's under the hood. We just need to understand what we can do with it.

A blockchain is not pretty, unless you are a coder with an eye for the beauty of keystroke commands. It looks like this (no need to do more than glance at this code, unless it intrigues you):

```
1.    # Create the blockchain and add the genesis block
2.        blockchain = [create_genesis block( )]
3.        previous_block = blockchain [0]
4.
5.    # How many blocks should we add to the chain
6.    #after the genesis block
7.        num_of_blocks_to_add = 17
8.
9.    #Add blocks to the chain
10.        for i in range(0, num_of_blocks_to_add):
11.        block_to_add = next_block(prevous_block)
12.        blockchain.append(block_to_add)
13.        previous_block = block_to_add
14.    # Spread the word!
15.        Print "Block #{} has been added to the
               blockchain!".format(block_to_add.index)
16.    Print "Hash: {}\n".format(block_to_add.hash)
```

This programming code translates into a usable interface when it is downloaded onto your device or computer.

Different companies or agencies can build apps in virtual layers on top of this code to make the information more accessible and useful to the people who access it on their computers, phones, or other devices. These apps can offer, for instance, a way to send money to your relatives. Or access to a marketplace of handmade clothing. Or entry to a social media platform, such as Steemit, where people earn money for making interesting social media posts, rather than giving their content away for free to behemoths such as Facebook.

Blockchains offer transparency while at the same time they offer a high degree of privacy. How is this possible?

The identity of a person (or machine) using a blockchain is hidden behind a string of code. Unless that person wants you to know who he or she is, you won't. However, the transactions conducted by that identity are transparent. In this way, you can determine that a transaction occurred, though you might not know exactly who made that transaction.

I say high degree of privacy, rather than total privacy, because it is possible for a good sleuth to figure out who you are based on an analysis of your transaction patterns. Companies such as Chainalysis and Elliptic do this all the time.

I arrived at the Assemblage, the coworking space and positive cult collective that's expanding rapidly in New York City and beyond. It's a friendly, trippy space on East Twenty-Fifth featuring a free Ayurvedic breakfast bar. "The Assemblage is a community of individuals who believe the world is at the verge of a collective conscious evolution transitioning from a society defined by individualism and separation into one of interconnectedness," is how the Assemblage describes itself. Who better to sponsor a panel discussion titled "AI, Blockchain, and the New Matriarchy"?

I entered the first floor, a vast high-ceilinged loft with stylish millennial furniture and a long, wide bar, and ordered a peppermint tea. The bartender smiled serenely. I sat on a stool and took in the gaggles of twenty- and thirtysomethings sitting on long couches, some of them snuggling each other, waiting for the panel to start. The five panelists sat on raised cushions below star-shaped chandeliers, their LED-lit dagger points directed at the floor.

I leaned back against a living moss wall as the moderator, integrative health and sexuality educator Anita Teresa Boeninger, opened the discussion by describing parallels between blockchain and matriarchal societies that tend to follow the nonhierarchical organizational structure called the heterarchy. She said these systems mirrored blockchain's distributed system with no apex point of power. "In the matriarchy, resources are distributed in different ways than in the patriarchal system," she said. "There is not the constant competition."

She instructed everyone to put their phones down and pretend instead that one of their palms was a phone. Behind her, on the wall, were photographs of smoke.

"Stroke your hands as if they were phones, touch your palms, really experience how you relate physically to your phone," she said.

I could feel the room loosening.

"Now, put your phones together on the floor so they can rub and talk to each other."

I heard a little gasp, but most people did as told. A clever way to get people off their screens to pay attention during the talk.

"Now let's get to our sexy panelists," she said.

I sensed a wave of anxiety roll across the room as everyone realized they were deviceless. Meanwhile, their phones were having all the fun, down on the floor getting it on.

Boeninger introduced Jesse Grushack, who works at ConsenSys, an innovative company that develops business and social good products on the Ethereum blockchain. The company has grown from 80 to more than 800 employees in just a few years, and has dozens of projects in the pipeline, ranging from a new journalism venture called Civil, to a local energy-trading platform called Grid+. Also on the panel were Maja Vujinovic, CEO of OGroup and an early figure in blockchain and other emerging technologies; Dr. Francesca Ferrando, a philosopher of the posthuman, with a background in gender research; and Alex Gordon-Brander, a leading designer of trading platforms who previously was chief business architect of ConsenSys.

Boeninger and the panelists were a good-looking, well-dressed bunch; the audience, on couches and cushions and deep chairs, was equally good-looking. Given that the room was also good-looking, this place had the aura of a wealthy Instagram fantasy.

Ferrando, an adjunct professor of philosophy at NYU, began by acknowledging "the nonhuman objects in the room—computers, phones, all these sacred lives. We manifested this space. We used technology. Technology is cocreating who we are. I want to share the vision of a world where there is no separation between us and technologies. Instead of saying, Us versus Them, we say, Us with Them. It is the end of patriarchy. The end of an era."

Cool, of course, but I felt a little vaped, as in vapidity. I didn't want to be cynical, feeling that I'd heard these thoughts fifty years ago, as a child in the seventies. I wanted to bask in the light of a future that was truly different.

Just like Boeninger, Ferrando was sprightly, with an asymmetrical look. She was intense and interesting. Even though her accent prevented me from understanding half of what she said, the half I could make out was excellent. Her ode to nonhuman objects kept me fascinated as I looked around the space—the planters, the piles of phones, the couches enveloping piles of scruffy young seekers. Still, I wanted blockchain talk.

Vujinovic spoke of her experience working in Africa, seeing firsthand the transformative nature of personal technology.

"With this"—she held up her phone—"you have a bank. And that bank empowers women in emerging market patriarchies."

That phone will also give these women access to whatever blockchains they desire, leveling power in a huge way, she said. Studies show that in places where women control their family's money, the whole society benefits. Blockchain, with its peer-to-peer payments and access to markets, is designed to make this happen.

This room was packed at least half full with women, a rare ratio at most blockchain events, but less surprising given that this event was, nominally, about the New Matriarchy. "My message to women," said Vujinovic: "Lean out, create the space you want. Don't wait to be invited."

Just then a tall, long-haired guy wearing a wide-brimmed Borsalino, black kurta, and elaborate kicks embellished with iridescent gold ribbon walked through the room, earbuds dangling and a laptop in his hand. A subtle patriarcher of the dandyish kind, the most insidious and indirect of all. The New Matriarchy parted to let him pass.

Hardly anyone had mentioned AI or blockchain, until Gordon-Brander joined in from his lotus position on top of a cushion. He was on the early side of middle-aged, with foppish hair and a suit, no tie. Clearly promatriarchal in manner, a Wes Anderson iteration.

"In Sumeria," he intoned, "money decayed over time. Your shekel became worthless with the years. Tech now, with cryptocurrency and blockchain, gives us the chance to rewrite the rules of money. I find that exciting. But still, 90 percent of bitcoin is in the hands of men."

The audience murmured in sad agreement.

"And that's probably not how we want to enter the posthuman world," he said.

Cheers from the crowd. Who, facing a subway ride back to Brooklyn on a cold spring night, wouldn't want to be posthuman?

Vujinovic, who had been an executive at GE, spoke up.

"Why is there all this hype around cryptocurrency? Because we all want more equity in the world. We still need to design the platform and the smart contracts with the intention of solving problems like poverty. We can't keep it under male-dominated thinking that creates tired consequences."

In other words, if we stuff all the new tech into the old world, everything will fail. The panel seemed to feel the paradigm shift should happen soon.

"In blockchain, time is compressed," said Gordon-Brander.

The panel ended. Furious networking began. I stepped outside to snowflakes swirling, my enlightenment on hold. The air bit my face but the snow was mesmerizing, a cosmic beauty that seemed designed to entrap me as I walked home.

I knew what a snowflake was. I'd seen the riveting images taken by Wilson "Snowflake" Bentley up in Vermont in the early twentieth century. He worked with an improvised camera and microscope he rigged up on his family's farm. Plenty of snow, but no electricity, no overt systems connecting him to the world at large. Even his family rejected his work. Yet those images are still the primary images we have of snow, shared for the last hundred years, visible to all of us at the push of a button.

Those images, no two crystals alike, made by hand on a farm in rural Vermont. The flakes fell on my dark jacket, disappearing in a moment, but somehow connecting me, through this distributed system of memory and reality, to that misunderstood man I would never have the chance to meet, Snowflake Bentley.

Ships Not Seen

This is the first information technology
I've encountered in my adult life that is
fundamentally difficult for otherwise intelligent
and highly capable people to understand.

—*Adam Greenfield*, Radical Technologies

Several early European visitors to the new world wrote that the local inhabitants seemed unable to "see" the explorers' large sailing ships offshore. Crew members with Christopher Columbus, in 1492, Ferdinand Magellan, in 1520, and James Cook, in 1770, all described this phenomenon, known as "ships not seen." Joseph Banks, a botanist who accompanied Cook, described one such instance:

> *The ship passed within a quarter of a mile of them and yet they scarce lifted their eyes from their employment; I was almost inclined to think that attentive to their business and deafened by the noise of the surf they neither saw nor heard her go past them. Not one was once observed to stop and look towards the ship; they pursued their way in all appearance entirely unmoved by the neighborhood of so remarkable an object as a ship must necessarily be to people who have never seen one.*

Cynics postulate that these stories are overwrought and infantilizing depictions of the local inhabitants. Yet I prefer another theory. Maybe the islanders could not see these ships because the vessels, which they'd never before faced or imagined, were completely outside their experience and perception.

My experiences with blockchain support this idea. In the last few years I've met many intelligent, seemingly well-informed people who find it impossible to imagine what blockchain is. They have never encountered any system resembling it, and they tend to go blind in the face of it. Blockchain is a ship not seen.

The science and technology of blockchain are challenging, especially for those of us who don't know much about computers and contracts. But lack of technical knowledge doesn't prevent us from being able to intuit or imagine the potential uses of blockchain in our society, or to use the dapps (the "d" is for distributed) built on top of it. Just as you are able to use the World Wide Web (WWW) without understanding how it, or the Internet beneath it, functions, you will eventually be able to seamlessly explore the benefits of blockchain from your devices.

You will probably never directly code a blockchain, just as you probably have never coded a website—even if you've "built" one on Squarespace or WordPress. In the next few years, smart programmers and engineers from IBM, Microsoft, ConsenSys, and dozens of other startups will have created friendly interfaces that allow you to set up whatever kind of chain works best for you or your company. Already, Amazon Web Services is offering templates for do-it-yourself blockchains. That's a pretty big tell.

It won't be long until using blockchain will be about as tough as watching a movie on Netflix.

The mind-expanding possibilities of blockchain, which is built on a distributed model, are well within reach. Again, by distributed, I mean that the system has no central authority, unlike almost every system we are accustomed to, but rather is run communally, with all the participants reaching a consensus on decisions. In a pure distributed system there is no hierarchy, no president, no father, no CEO. When all participants are valued, the rules of business, creativity, and society change.

How do they change? We need to figure that out.

The potential of distributed systems, what they might mean for government, creativity, equality, and access to money and business opportunities for all, is so exciting, and new, that the future is hard to fathom.

To understand blockchain, feel blockchain.

Imagine yourself on a beach. The sun beats down as a foggy breeze off the water cools you. Thousands of other beachgoers share the sand: Tanzanian mountain climbers, Uruguayan stock traders, poor farmers from Mato Grosso, a private jet pilot from Long Island, three English eel fisherfolk, a French art handler, and a movie star from Silver Lake with 549,368 followers. None of you know each other, yet all of you are connected. Separately, each one of you reached the decision today to go to this beach and enjoy the sun. Each of you can confirm that this happened. The collective truth is established. That's the blockchain.

This is also the blockchain: You are riding in *Car and Driver*'s best vehicle of 2025. It is shaped like a three-sided onigiri, and the entire surface, including the wraparound windshield, is composed of highly efficient solar panels that store electricity in a battery that sits where the engines sat in twentieth-century cars. The bottom of your car has a conducting strip with a smart meter connected wirelessly to a blockchain. AI largely drives the car for you, and when you arrive at a stoplight, your car wirelessly downloads your excess power to the grid. Automated digital smart contracts keep track of everything, including who bought your power, how much they paid you, and where that money is right now. The car can pass through innumerable grid boundaries, from Duke Energy to Uncle Joe's microgrid and on to the Central Virginia Electric Cooperative, with all the transactions handled by smart contracts and applied to your account. Your car will create and sell electricity all day long, and at night, when solar panels don't work and the wind is low, digital agents for the grid will call upon your car to release the electricity stored in its battery to supplement the overall electricity supply. The smart contract will be sure to leave plenty of juice in the car for you to get to work in the morning, and you'll have earned some money on the side.

———

The Internet, as we know it, becomes less important. It's the Internet of information. The new, distributed Internet will grow out of this old Internet, coexisting as the Internet of value and truth, thanks to blockchain. At first, the new Internet will run on the old Internet, like an egret on the back of a water buffalo. Just as the egret serves a purpose by eating parasitic ticks off the buffalo's hide, Internet New removes lots of mistrust from Internet Old. As the buffalo provides food for the egret, Internet Old offers support for Internet New. Meanwhile, Internet New flies upward, into the new age of enlightenment.

Eventually, the new Internet will run off its own devices scattered in blockchains all around the world. Empty spaces in the memory of phones, cars, mainframes, and other devices, perhaps even in our brains, will be occupied by this new Internet. Melanie Swan, a technology theorist at Purdue University, calls it the crypto enlightenment. Call it what you wish—it has begun.

It takes an almost spiritual leap to conceptualize how blockchain behavior will affect human behavior, business, sustainability, and other important systems. We see varieties of distributed and decentralized systems in nature, and I find these examples help me understand human and machine patterns.

For enlightenment, of course, I turn to YouTube. (And I recommend that you, also, watch YouTube videos of distributed natural systems as you think about blockchain.)

Most inspiring, perhaps, are the videos of swirling fish (one, called Tuna Tornado, is well worth a watch) and the fantastic flight patterns of starling flocks, also known as murmurations. It's believed the murmurations are composed of hundreds of nodes, each made up of seven birds. Each bird needs to be aware only of what the other six birds are doing. Yet every movement they make affects, and is affected by, all the other nodes. When a predatory bird appears near the murmuration, just a few nodes will break off to deter it from attacking. Similar systems are at play with schools of fish that form moving, slippery structures to defend against sharks and other predators.

Ant colonies function without any central control, with members of the group signaling behavior changes and responses by secreting chemicals. Swarms of bees communicate in distributed ways, improving the lot of the group. These insect groups might have one leader, but the communication among the masses is indirect, leaderless, based on chemical cues or flight patterns. In less visible ways, cells organize into networks. Neurons collectively induce sensations, such as pain.

We humans are part of this grand system of nature and creation. I believe we are genetically inclined toward decentralized and distributed systems. We just need to open ourselves to the potential that they offer.

Blockchain is not a moral agent, and despite the hype, the technology has no intrinsic ability to improve our society. Rather, it is a soul-free tech platform on which to build apps that can influence most aspects of our lives. Blockchain's benefits depend on how we use it. It is so powerful, and so novel, that it gives us the opportunity to think in new ways, to act more collaboratively and challenge our assumptions of hierarchy. It offers us the option of technology as a force for good and for commerce at the same time. Blockchain has the potential to transform broad categories of industry, science, government, and society—even the earth and its climate. It invites big thinking.

It also demands that as we imagine its incredible potential, we keep in mind that this technology is in its infancy. Like any infant, it is shaped by its environment, and the people it comes up against.

We've seen other technologies, including the Internet and the subsequent WWW, veer sharply from their initial utopian promise. Blockchain is no less susceptible. Already, Wall Street firms and banks, threatened by the encroachment of cryptocurrencies and blockchain finance models, are working to incorporate it into their own ways of doing business. In the summer of 2018, Mastercard got a patent for a method of speeding up blockchain transactions. Leading investment banks are flirting with starting cryptocurrency trading desks, because they stand to benefit enormously if blockchain technology makes intermediaries, which charge money to transfer money, obsolete. Social media giants are wary of the tech's threat to their control of personal data, and Facebook has launched a large investigation into the uses of blockchain, which threatens its very business model. Certainly, these corporate titans will seek to dominate the space.

On the other hand, cryptoanarchists, many of whom subscribe to an extreme type of libertarianism, see the cryptocurrencies associated with blockchain as tools for a future free of government or societal control. They envision the end of banks. They believe cryptocurrency will supplant fiat currency, which is government-issued money that is not backed by anything tangible, such as gold. (That would include the U.S. dollar.) And then there are the United Nations, the Rockefeller Foundation, and the World Economic Forum, which are developing blockchain technology to help disadvantaged farmers, people with no access to banks, and disenfranchised voters, among other social projects.

The future of this tech will be determined by the ways we exploit it, and also share it with each other, because blockchain is the most collaborative of systems. As we explore blockchain's uses, and create applications, we must remember that things might not unfold as we expect they will. Vigilance is important if blockchain is to fulfill its best promises.

Can you conjure up a Segway in your mind? The year before the Segway was released to the public, in late 2001, everyone was sure this crazy new invention was going to change the world for the better, in a big way. Supposedly, Steve Jobs predicted Segways would be as influential as personal computers. A prominent venture capitalist said the invention was more important than the Internet. Then the two-wheeled scooters with intuitive self-balancing mechanisms came out, and most people went, "Wow, that's it?" Nobody really cared. Now mall cops and tour groups use them sometimes, but do you know anyone who admits to owning one?

Maybe blockchain is the Segway of our day. Maybe the hype is over the top.

Even though hardly anyone can tell you what blockchain is, a lot of people hate it. The most common challenge, usually from a guy wagging his finger in your face, is: "Well, what actual viable business use of blockchain can you point to?" That's followed by a self-satisfied smirk.

I usually point to cryptocurrency. What could be more successful than bitcoin, which had a market cap of about $60 billion in December 2018. Even accounting for its downward spiral from a market cap of $220 billion in December 2017, it's a powerful market force.

Bitcoin is a world-altering tech, and blockchain makes it possible.[*] This digital coin has produced a large number of obnoxious young bohemian millionaires, and is also being used as a payment system by about one in five banks on earth. It's used by criminals to transfer money without detection, and by aid organizations in lower-income nations. Personally, I find the aesthetics and culture of bitcoin off-putting. I hate the name. Yet, there's no denying its impact on our business and social evolution.

---

[*] Yes, there are huge environmental problems associated with bitcoin and other "proof of work" systems. We address these, and solutions to the problems they cause, later in the book.

It was a hot September afternoon in 1993, when a URL plastered on the side of a passing bus lit up my senses. I had no idea what it meant, but I knew it was a code I wanted to crack. As I walked through the crowds along Fifth Avenue, headed to do research amid the horrible smells, endless card catalogs, and scratchy microfiches at the New York Public Library, I pondered: HTTP:?

At first, the stream of nonsensical black letters on a yellow background were jolting. Clearly, the ad that had transfixed me was meant to convey something significant, but I had no idea what. I'd been away in the remote Andes mountains for three months, and this ad taunted me with the idea that I'd missed something big. What could WWW mean? I'd used computers regularly since paying $3,500 for the first model Macintosh, in 1984, but the web was still a newborn.

Of course, within five years those letters, http://www, had transformed my work life as a journalist, and I've hardly been to the library since.

We are at a similar societal shift right now, only this time the codes don't appear on the sides of buses. In fact, it's unlikely that you've seen them. They look like this:

36ebee7dd9c5ff23a24f20334f16c27ece
718790a2871221cffe4508a6f1c581

This inelegant string of numbers is called a hash, and it's a symbol of our new reality. The hash represents a land title, photograph, novel, money-wire, Sam Cooke song, or grocery list—whatever info a person wants to keep track of on a blockchain.

These cryptographic streams are the foundation of a new paradigm called the distributed age. In terms of technology, we're where the Internet was in 1995: a place where nerds, money people, and creative entrepreneurs know that something big is happening, while the rest of the world just shops for Beanie Babies at the mall.

For a long time, our habits, politics, finances, and arts have been shaped by centralized thinking—we give ourselves over to authority figures or institutions to tell us what is right or wrong. We request validation from these central authorities, and they dispense it as they see fit. For instance, we look to banks to confirm the value of our transactions, to museums to affirm the value and provenance of art, and to political party leaders to guide us forward.

———————————

We are very comfortable with this situation. Take a cursory look at your life and you'll see that you put your utmost trust in a variety of institutions and systems, including:

- The utility company
- The insurance company
- That stranger you met online you're hoping to cook dinner for and sleep with tonight
- Your new landlord
- The rare book dealer trying to unload a first edition Harry Potter

Why do we trust these institutions? Because they are more powerful than us in some way. Also because, most of the time, we have little choice. While institutions are famously capable of hanging on to power, it's possible that those days are ending.

Embracing Distributed Systems

When we try to pick anything out by itself, we find it
hitched to everything else in the universe.

—*John Muir*

The difficulty many of us have truly understanding distributed systems speaks to our conditioning. For at least ten millennia, humans have lived in increasingly hierarchical groups, defined by pecking orders and leaders. It's been necessary for people to learn their place. If they don't want to stay in their place, then they have to fight their way up the ladder. The concept of hierarchy starts with the family and expands ever outward to the local community and the larger community of the state, region, or world, based on social status, wealth, and power. It's true in businesses, which are controlled by small, elite groups, and it's what causes the most agitation on the global stage, as when major players, such as China and the United States, jockey for hierarchical dominance, and less powerful states, such as North Korea, assert their own power.

Philosophers, anthropologists, and others say it wasn't always this way. Some evidence supports the idea that hunter-gatherer societies were egalitarian, and put a lot of effort into maintaining equality among members of the group. According to James Woodbury, of the London School of Economics, the egalitarian qualities arose because of several factors: Hunter-gatherers had immediate, direct access to meat, fruit, and nuts that the group gathered, which limited outside control; goods could move among people without any of them feeling beholden to another; and people weren't unduly dependent on others, as all contributed to the outcome for their group.

Apparently, say the anthropologists, hunter-gatherer societies did not have chiefs, bosses, or royalty. Women and men both had important functions. People cooperated, because that was the best way to get something to eat in a world where hunting and gathering was a daily, time-consuming task. Then hunter-gatherers figured out how to raise crops, about 12,000 years ago. Suddenly there were goods to manage, and other tasks to accomplish, in ways that changed these societies forever. As human groups grew, and used more resources, they needed to take over other territories to find enough animals to hunt, and enough good land to cultivate. This in turn led to the need for people who could manage the resources, which were now distributed unequally, breaking the egalitarian structure. Suddenly, the group was divided by rankings based on prestige, wealth, and power.

Stanford University researchers ran a simulation that showed that this type of unequal access to resources weakens, rather than strengthens, groups. Yet at the same time, according to the models, the inequality enlarges the groups, by causing them to spread farther in search of food and other necessities. This leads to conflicts with other groups. When unequal groups conquer neighboring egalitarian societies, inequality spreads even further.

Our current unequal societies led by dominant hierarchies are a direct result of the transition to agriculture thousands of years ago. This state of society works well for many of us, and not so well for most of the people in the world, who struggle to have good food, health care, education, and even such basic necessities as clean water and sanitation.

Maybe it's time to look at the systems that control much of our behavior and try something new.

Blockchain tech is called distributed tech because every computer or other device on the network participates in its functioning. The devices, called nodes, can be right next to each other, or on different continents. The nodes have to form a consensus about every block created and time-stamp it on the chain.

There are many ways to form this consensus, ranging from souped-up computers competing to solve algorithmically generated "problems," in a process called mining, to people proving their worthiness to verify consensus based on the number of cryptographic coins they hold—known as proof of stake.

At first glance, this distributed nature made no sense at all to me, although at the same time it gave me a little illicit thrill. It just seemed crazy to have a level playing field, with no obvious leader setting the ethos of the organization. It upended everything I knew.

Yet, I've come to realize that this is the most important innovation of the chains, the aspect that is going to transform our future, if we let it. In a distributed system, no one node makes all the decisions. What's more, since each node is aware of what the other nodes are up to on the blockchain, there's a great chance that they will collaborate in more fruitful ways than in a top-down system.

Distributed systems have the ability to give power to the powerless. They also let the powerful flourish when they take advantage of the creative thinking brought on by collaboration.

In general, most of us live lives that are dominated top-down by those who have expertise, power, wealth, social standing, or some other means of control. This system is so ingrained in our societies that most of the time we don't even notice. Those among us who "have trouble with authority" quickly learn to pick our battles. If we don't, we suffer.

The dominant force might be the head of a family, or it might be a chosen leader, such as a president. It might be a corporation, government, or religion. It might be someone who has taken the time to absorb more knowledge about a subject than anyone else—or at least pretends to have done so. These forces make decisions for us and serve as gatekeepers to the truth. This authority comes in two basic forms: centralized and decentralized.

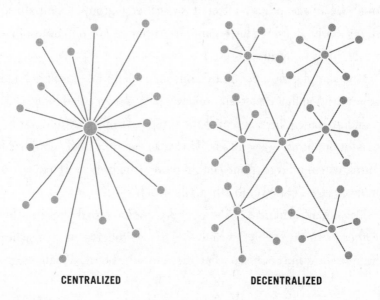

CENTRALIZED   DECENTRALIZED

Centralized systems flow from the dominant node to all the lesser nodes. Let's say you are a small-scale palm oil producer in rural Indonesia. There aren't many of you left, as the industry is dominated by giant plantations, but those few of you remaining sell your crop to a local aggregator who has connections to the larger market. Let's say this aggregator offers you 14,000 Indonesian rupia (about one dollar) for each large bag of palm oil nuts you harvest. There is no negotiating on the price, and no way for you to ask for more money, even though your harvest was sustainably raised (unlike those at the large plantations). Your choice is limited to two options: sell, or don't sell. That is an example of centralized authority, and it's in the hands of the aggregator. In effect, you have no flexibility and almost no choice.

Decentralized systems also have a central power, but that power gives power to other nodes, which in turn dominate the bulk of participants. This system can feel less direct, but also more oppressive, though we all accept it. In effect, it is a scattered group of centralized systems dominated by a large centralized power. Corporations often operate this way, as do schools.

For example, my older brother once objected when his fifth-grade teacher insisted that Baja California was part of the United States. For his intemperance, he was sent to the principal's office. There was no space for opinions, or even facts. That is an example of decentralized control, wherein there is one central power (the principal) that farms out lower levels of power to others (the teacher).

Despite all evidence to the contrary, the principal kept him for the afternoon. My brother's twenty-five fellow students, in their sixties now, must be quite confused when they cross the border at San Diego.

The terms "decentralized" and "distributed" are often boiled in the same pot. I think that each deserves its own pot. Decentralized systems are nonhierarchical groupings of nodes that are connected to each other. For instance, the terror cells that make up Al Qaeda. Or the football teams in the NFL. Distributed systems are radical decentralized systems. The meetings under the Alcoholics Anonymous umbrella, for instance, are independent, self-funding, and can be started by anyone, yet belong, in a very real sense, to a larger organization.

Distributed systems are so radical that they are hard to imagine, let alone maintain, in most cultures. Decentralized systems seem to be more manageable, and still retain many of the benefits of their radical cousin. Blockchain is, by design, a distributed system. However, decentralized systems use and benefit greatly from its structure.

Many people look at a diagram of a distributed blockchain network for the first time and judge it as weak. There is no center. No dominant node. With everything in balance, who's going to make the decisions? How will anything get done? All perfectly sensible responses.

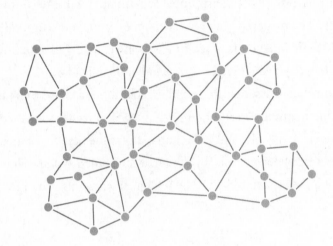

It's a pretty reasonable fear, and one I still feel. I've spent a lot of time wrapping my head around the distributed system concept. As I've said, after much pondering, I see the beauty of it.

The illustration above is like a web. If one part of it were to be destroyed, the rest of it could continue to connect. All of the nodes are part of a greater whole. Each node can contribute. In effect, all of these nodes, or participants in the blockchain, work together to form one large computer. As part of the system, each node can see what the other nodes are up to. This transparency should enhance creativity, as people who once were off the radar can now think about and participate in solutions and suggestions.

Now that I see the beauty of distributed systems, the centralized structures from the past feel antiquated, lethargic. Pondering them I feel stifled. I don't want a big boss telling me what to watch, think, or do. I'd rather be able to make my own contributions to the outcome. This was the promise of democracy, communism, and socialism. The idea that the people would have the final voice. Democracy and socialism are still hanging on by a strong thread, but clearly communism never lived up to its billing. Maybe distributed tech, fueled by blockchain, will lead us to some new, more egalitarian options—what the philosopher of blockchain, Melanie Swan, calls the next enlightenment.

The more participants in a distributed network, the better it works.
An influential electrical engineer named Robert Metcalfe first demonstrated what has come to be known as Metcalfe's Law in 1980, regarding phone networks: Two phones can only make one connection, with each other. Five phones, however, can make ten connections. And if you have twelve phones, you can make sixty-six connections. The number of connections rises exponentially as the number of participants grows. This "law" is frequently challenged and might best be described as a rule of thumb. Many factors affect the growth in connectivity. In general, the more connections a network can make, the more secure it is. Also, the more potential there is for creativity and transactions.

Curiously, Metcalfe invented Ethernet, which preceded the Internet, and he's notorious for predicting, in 1995, the total collapse of the Internet within one year. He promised to eat his words if it didn't happen. At the 1997 World Wide Web Conference, he put a printed copy of his prediction into a blender with some water, whipped it up, and then drank it in front of the audience. He later also predicted a quick end to the wireless craze. There's always room for characters.

The technology encourages an unprecedented way of organizing, creating, and sharing value. When Columbus returned to Spain with goods and slaves from the new world, he opened a door not only to material wealth, but to the future. Some people saw the possibilities and began to explore and exploit the new lands. This impulse led to great harm as well as great achievements. Other people grew fearful and retreated.

Blockchain is our new world, and our reactions to it will determine our place in the future. Columbus dove in, and widespread annihilation of cultures quickly followed. Technology often follows that same trajectory. Perhaps this time we can go forward and explore this new world without destroying the best of what already exists.

Gavin Wood, a lanky Englishman who has been involved with the idea of distributed ledger systems since the early 2000s, before bitcoin existed, describes a blockchain simply as "a computer shared by everyone." Because the transactions recorded on the blockchain are shared with every computer connected to that chain, the organic cotton grower in rural Vietnam is as much a steward of the chain as the London fashion executive who sells that cotton in a refined form. In this regard, blockchain resembles the fabled "commons" of yore—the town green where everyone could graze their sheep. Unlike this prosaic shared greenspace, blockchains do not lead to the "tragedy of the commons" that every Econ 101 student studies. That tragedy occurs when all the members of the community try to exploit the grass of the commons to their personal advantage, with no consideration given to conserving grass so their neighbors' sheep can also graze.

On the blockchain, in theory (it will not be absolutely proven until blockchains are in wider use), it's nearly impossible for one participant to manipulate the chain to his or her greater advantage, unless that individual gains majority control of the computing power on the chain that certifies transactions. If that happens the invader can block others from acting, and even steal cryptocurrency.

For years, the 51 percent attack was a threat, like Sasquatch, that many feared, but none encountered. However, since 2017, a few of the smaller altcoin blockchains have successfully been attacked. Because the smaller systems have fewer nodes that certify transactions, it's relatively easier (though certainly not easy) to buy or rent enough computer power to overwhelm them. Larger chains remain impervious, although many chains, such as bitcoin, are said to have their active nodes concentrated in the hands of a very few people. Many people are working on security solutions.

Since anyone with a computing device can become a node on a public chain, the barriers to entry fall, making it easier to join the blockchain commons than it is to enter many of the "rooms" now living on the Internet. You won't need to ask for access to a public chain. Even "private" blockchains, which control access to information in order to maintain privacy, will offer an unprecedented ease of access, in time.

The distinct functions of distributed networks will alter our minds as blockchain expands. Never in history have humans made significant decisions, repeatedly, in a distributed way. With blockchains, we are becoming a hive.

Back in 1981, when the American pop artist James Rosenquist created a series of seven prints titled *High Technology and Mysticism: A Meeting Point (Seven Works)*, the Internet was still birthing from ARPANET, and rare digital art was far in the future. Distributed thinking was in the air, for sure, though blockchain and its seeds were not yet in evidence. Rosenquist was celebrated for startling, oddly joyous images of handguns on candy-colored backgrounds, massive canvases assaulted with vividly colored lipsticks, fighter jets, spaghetti in sauce, and John F. Kennedy. The images were overlapped and fragmented and compelling enough that as a teenager I stared at his canvas *F-111* for more than twenty minutes the first time I saw it.

In contrast to the urbane political nature of these canvases, the *High Technology and Mysticism* lithographs are deep meditations about technology and humanity. Distributed systems of various sorts, represented in soft lines and hard numbers and circles, like nodes, are layered over distorted images of people and animals. At the time, computers were just beginning to become usable for nonexperts, and it seems Rosenquist sensed the impending digital age, as well as the connection of mystics to distributed systems.

Mystics have always investigated nonhierarchical worlds, seeking wisdom and connection to God. Some Sufis cross dimensions in meditation to communicate with the beyond—though to them, this mysterious beyond is just "right here, always." The communication between spirit, ancestor, self, and neighbor is immediate and always present. Christian saints and martyrs, such as Teresa of Avila, have expressed a similar connection with "nodes" in other places.

I remember once traveling to the Ecuadorian Amazon with a highland shaman named Ernesto, to join other shamans gathering to heal a Shuar wise man who had taken ill in his village. It was an epic trek by small plane, canoe, and foot to reach the village, and the esteemed shaman seemed near death, lying on his side on the ground, a small fire warming him.

Several of the lowland shamans had gathered the vines and other materials needed to prepare their sacred ayahuasca, a powerful potion they were going to take in order to travel into another dimension to seek help and wisdom to heal their sick brother. As night came down, the shamans gathered in a circle and chanted, a few playing drums, each wearing the traditional clothing of his group. They were all men. The ayahuasca, dark and bitter, was served in small gourds. The shamans drank. (I declined.) Time passed and some of them vomited in the forest at the edge of the clearing. The only light was from the fire and a lantern hanging from a pole.

An hour or two passed—I do not know—as I watched these shamans chant, pray, and sing around their sick colleague, who still lay on his side.

Suddenly, a fluorescent green line appeared along the bodies of several of the men. It was radiant with periodic dots—nodes—like the lines of a Keith Haring painting, traveling geometrically in the darkness, a marching dotted outline of the shamans' bodies. I, who had not taken any drugs, was absorbed by the light, the movement, of the vibrating lines, constantly replenished.

Looking at the James Rosenquist lithographs I see these lines again. These distributed systems are the same systems that the artist used in his series. They are the same patterns we see in the distributed systems of blockchain. The patterns we see in groves of redwoods and aspens, connected and communicating through underground roots. The huge starling flocks that make startling patterns in the evening sky—one moves, touches a wing to the next pod, and the geometry of the entire murmuration changes, pod by pod.

II
HOW IT WORKS

The Chains

The business value-add of blockchain will grow to slightly more than $176 billion by 2025, and then it will exceed $3.1 trillion by 2030.

—From a forecast by analysts at Gartner Consulting

The first and still most famous blockchain was designed in 2008 to support bitcoin, the digital cryptocurrency invented by a reclusive and very mysterious cyber-agent who has never appeared in public. At first, nobody thought much about blockchain as a tool with broad applications. It was just code that made bitcoin possible. But soon people began to realize the platform might support amazing innovations. There are many other blockchains now, for other currencies called altcoins, and also for fashion companies, Walmart, artists, the government of Dubai, collectors of cat images, protectors of endangered redwoods, derivatives traders, and others, with many more on the way. As many as we want to create.

As a group, all the blockchains are called distributed ledger technologies.

Several platforms dominate the blockchain space, including the one used for bitcoin. Hyperledger, a platform based on Linux code, is used by IBM's enterprise blockchain team. It does not depend on coins, and the chains are often "closed" or confined to certain business teams, so they can maintain their privacy. Ethereum is another, and uses coins called ether. Ethereum is probably the "cooler" of these dominant platforms, although Hyperledger is favored by an abundance of more established businesses. There's no reason these and other platforms can't coexist. A consortium is now trying to set standards to allow the various platforms to communicate seamlessly. There's a lot riding on working together.

Generally, actual content isn't stored on a chain. Your digital documents, music, images, contracts, and other transactional items are "hashed" into a cryptographic code that represents the content, but contains only a string of sixty-four letters and numbers (if it is what's called a 256 hash—other hash functions will have different characters). The original documents or visual files are stored off chain. Think of the hash as a thumbprint for each item.

To create a hash, you enter data, whether it be a photograph, a scan of a painting, a PDF, a Word document, or a financial transaction, into a program called a hash generator. To try this yourself, Google "256 hash," and a generator will appear. Type a few words into the window.

For instance, if you type abc (in lowercase), you will get this hash:

ba7816bf8f01cfea414140de5dae2223b00361a
396177a9cb410ff61f20015ad

You can do this anywhere in the world, on any device, and it will always come up with the same hash. This is true for any collection of digital information. If you input the exact same data, you will generate the exact same hash. That means that the same manuscript, hashed at different times in different corners of the earth, will give you the same sixty-four-letter code.

But typing ABC (in caps) will generate this hash:

b5d4045c3f466fa91fe2cc6abe79232a1a
57cdf104f7a26e716e0a1e2789df78

This hash is different from the hash for abc in lowercase. That ensures trust. To check the validity of any data, such as a digital land title, just hash the document in question. If the resulting hash is not the same as the original recorded on the blockchain, then you know the land title or other document has been altered. Even changing just one character or space in a document will result in a new, entirely different, hash.

This blockchain operating system can live on any type of computing device, including a refrigerator, gas meter, or electric car, when it is connected to something called the Internet of Things (IoT). The concept of the IoT is that as we move into the future, more and more inanimate objects will be configured to connect with the Internet, to make it easier to know, for instance, when your refrigerator needs to be stocked with food, or when your electric vehicle (EV) batteries are storing extra power that could be sold to the electric grid. Blockchain technology is perfectly designed to help these smart objects communicate with each other. It even enables them to pay each other for services, using automated contracts without human intervention.

In a sense, all of these nodes in a blockchain—computers, phones, refrigerators, EVs—link together on the Internet to form an even larger computing device. The blockchain software lives in all the nooks and crannies of this large device. It's the Milky Way of connected computers.

While all of this sounds very complex, in truth you can easily access a blockchain through your phone and take advantage of smart contracts to record financial stuff, artwork, legal processes, the materials used in making a product, your personal ID, and tons of other information. Once the data are entered into a block, and the block is electronically sealed, it can't be changed.* Whatever you have recorded becomes part of a permanent record.† For instance, you could record your idea for a new movie. The exact words that you use will be represented in the coded hash, on the chain, within a time-stamped block that contains data from other people's transactions. No one could ever question that the work originated with you. It will always be there for the world to see. If you update your script at a later point, you just enter that new version in a new block. The two versions will reference each other. The record is duplicated and updated on all the nodes in the chain, easily visible and indisputably recorded without the need for intermediaries, such as the authors' guild or a copyright lawyer.

This unchangeable nature is already being harnessed by innovative entrepreneurs who are having an impact on health care, finance, global money transfers, property titles, sustainable behavior, dating, literature, supply chains, frequent flyer miles, food safety, and much more. It has the potential to make us safer, richer, more efficient, healthier, and more equal. Let's hope we let it.

---

* However, developers are now building ways to make some blockchains "editable" for highly specific purposes.

† Similar to the dreaded "permanent record" my junior high principal warned would follow me throughout my life if I didn't behave better.

On a well-known episode of the television show *Portlandia*, the characters Peter and Nance consider ordering a roast chicken, but first must be convinced of its bona fides. Their server says the bird in question is a heritage breed, woodland-raised chicken that was fattened on a diet of sheep's milk, soy, and hazelnut. She brings Peter and Nance a folder of paperwork about the chicken, which was named Colin before it was butchered and plucked for dinner. Did the chicken have friends? Peter asks. Like a friend it could put its wing around when it wanted to? What is the farmer like? The waitress, chagrined, says she does not know that level of intimate detail about Colin. In the end, Peter and Nance leave their table and drive to the farm to have a look for themselves. With blockchain, they could have just checked their phones. (Ag company Cargill has already experimented with using blockchain tech to help people who bought Honeysuckle White Thanksgiving turkeys track their provenance.)

Blockchain helps humans accomplish many important tasks with applications called dapps. They can be built in various layers on top of a blockchain to serve many purposes. For instance, a dapp could look like the website of a real estate company, with photos of listings for sale in a blockchain-based system. In this way, dapps make using blockchain intuitive for those of us who aren't technocrats, much the way that the WWW made using the Internet intuitive and breezy, rather than an exercise in coding. Well-constructed dapps have the potential to serve society by:

- Enabling direct trade between people, without the need for intermediaries, such as brokers. Sometimes called peer-to-peer connection, this trait could encourage a true sharing economy that would obviate the need for massive centralized corporations like Airbnb.

- Fractionalizing the ownership of physical assets, such as rental buildings and Picassos, into small, easily traded units of value. This makes it possible for anyone to share in these economies that have always been within the purview of the elite.

- Preserving the truth—historical references will be immutable, in the form that they are entered on the chain (the veracity of each fact will need to be certified before it is entered, and that proof could accompany the information for the rest of time).

- Allowing people to exchange money with each other at little or no cost, with total confidence.

- Letting people own their own data, which they can sell, rather than giving it for free to such giants as Facebook, or just keep private.
- Storing personal identification that can't be challenged.
- Giving businesses the ability to monitor supply chains and make them transparent.
- Increasing business and government efficiency by removing the need for bureaucratic interference.
- Preserving ownership of intellectual and physical property.
- Allowing an economy of sustainable capitalism that will benefit the planet.

The potential appears to be infinite.

Appears to be. This is a key phrase when it comes to blockchain. When I first came across a reference to blockchain's potential to change the world, I was immediately hooked on this tech that was being touted as a cure for everything from financial inequality to the nuclear arms race. My deep dive since then has taught me that blockchain is not quite as simple or as immediately transformative as we might wish.

As the clamor around blockchain has increased, so have the warnings. The tech might not be as scalable as people had hoped. It might not be able to process transactions quickly enough. It might not be as secure and immutable as most people still say it is. It might be usurped by corporations that control future innovations and keep them private. Some of the methods for securing blocks use vast amounts of electricity, and that won't be tenable. The cryptocurrencies are Ponzi schemes, say some. And perhaps most important, as blockchain enters the mainstream, people are discovering that it's twice as hard to understand as they hoped it would be.

The smart contract is one feature of blockchains that inspires both utopian thinking and fear that blockchain will never live up to the hype. I cannot predict with any accuracy whether these algorithmic, self-executing contracts will prove to be reliable or simple enough to help machines communicate with each other, pay people for services rendered, and launch a new period of business and social innovation. But I do hope they will.

I've never been much of one for contracts, beyond a handshake and a strong gaze into the other person's eyes. I've seen others break contracts with impunity, and I've signed contracts that all parties ignored completely. I've seen contracts enforced unreasonably, such as the time an alarm company insisted I keep paying for service even though I'd sold the house. I like to go with my gut and trust whomever I have an agreement with, contract or not.

So why am I so excited about smart contracts?

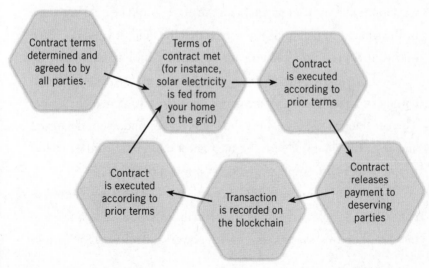

First of all, the name is great, especially compared to most block-chain-related names. Second, the contracts turn the straightforward blockchain ledger into a revolutionary tool for automating agreements, so that business, personal, and societal activities aren't slowed down by bureaucratic logjams. Plus, in theory, they are completely trustworthy, because once they've been programmed, no humans, courts, or other wobbly institutions need be involved in their execution.

"In theory" is another apt phrase when it comes to the optimism around blockchain. It rightly suggests infallibility without guaranteeing it. In practice, there are still security concerns with smart contracts, and one 2018 study found a 3 percent failure rate for smart contracts on the Ethereum chain, which in some cases could lead to a loss of money. That's significant. Yet many people are working on these issues. If the contracts prove to be reliable, as many people believe (though there are serious dissenters), they could make many giant centralized companies, such as Airbnb, obsolete.

For instance, if you were to rent an adobe casita in Santa Fe for a weeklong getaway, your payment for the place would be held in escrow by the smart contract until you agreed that the landlord and the place met your expectations. Then your payment would automatically be released. If you overstayed the agreed-upon rental period, the smart contract would either negotiate new terms or lock you out of the house. The contract would also notify the housekeeper upon your departure and reset the door code for the next guest. No credit cards, just low payment fees and simple peer-to-peer interactions.

A smart contract can read a meter that tells it when a solar power generator has released enough electricity to power someone's car; monitor that car's downloading of that power; and release money from the car's account back to the owner of the solar panels, all without a pump jockey or a credit card.

Let's say I'm the owner of that coltan mine in the DRC that I mentioned in the intro. I supply you with fifty kilograms, which you plan to sell to Apple to use in its smartphones. Once you accept the mineral, your money, in escrow, will be released to me. But only if the coltan is certified as sustainably extracted, with no harm to gorillas or other vital forest elements. Additionally, a smart contract for Apple, further up the supply chain, sees the volume of coltan has entered the market, and another smart contract will automatically increase Apple's rate of production. Meanwhile, the government of the DRC is able to track the origin of the mineral, and to follow its flow. No more need for cumbersome old systems that rely on the flow of actual paper documents, or digital files, from person to person up the chain.

There is a certain cruelty to the finality and certainty of smart contracts that will tax the patience of many people in our society (including me). While terms of a smart contract can be negotiated, they are also rigidly mechanical. You can't fudge—staying an extra few hours after checkout time, for instance, if it's not allowed by the contract. While to my mind this limits human freedom, it has a huge plus. When you bind yourself with a clear, automated contract, you know exactly what to expect, and also what is expected of you. The contracts set clear boundaries. This could keep you out of court and prevent hard feelings.

While this might reduce the need for litigators, it would definitely increase the need for contract attorneys to create the smart contracts before setting them loose on a blockchain.

At the moment, most business in the world is conducted with great secrecy. Blockchain tech offers the possibility of radical transparency, so that all participants in the chain (everyone in the world, if desired) can have access to every transaction on the chain. That means, again, that a small-scale palm oil producer in the Ivory Coast would be able to track the flow of her farm's oil to the regional aggregator, the processor, the transporter, the shipper, the customs broker in the United States, the packager in Philadelphia, and then the grocery store in Manhattan that sells it. The person who buys a jar of organic red palm oil off the shelf could also access the chain, to ensure that the farmer who grew it didn't use pesticides. This openness empowers the producer, way down the chain, to demand fair prices. It also gives everyone along the chain access to data they might use to improve the product, sales, or distribution and marketing. The supply chain becomes a partnership chain of shared data.

There are both private blockchains and public blockchains, and they operate pretty much as they sound. Purists and crytpoanarchists and other blockchain fanatics generally believe that private chains violate the principles of open transparency that can make blockchain so radical. Public chains are like public schools—open to everyone. Private chains are like private eating clubs at Ivy League universities—you must be invited to join. Private chains are mainly used for what businesspeople call "enterprise." If you've got a big old corporation, and you want to keep your supply chain private, then you'll put it on a permissioned chain. IBM does a tremendous amount of work with these chains, for a lot of big operations, using the Hyperledger Fabric framework. The main drawback to permissioned chains is similar to the problems that the private eating clubs have: You don't get much interaction with creative minds from outside your friend group, since none of them come to eat with you. With private chains, everyone preaches to the choir.

For good reason, this appeals to some companies that would like to keep their supply chain secret; to health care operations that want to give patients maximum privacy; and to others who haven't yet embraced the modern movement toward transparency.

The great beauty of public blockchains is that they are transparent, and open to anyone. That means that all the transactions are visible, even though the identity of the participants might be hidden behind the cryptographic code of the public keys that let them initiate and receive transactions. It all depends on what the node wants to release about itself.

As blockchain is adopted by more and more corporations, some people fear that it will lose its anarchic, free spirit. This is a very reasonable fear, and one that has been realized before, with other technologies such as the Internet and WWW, which in the commercial realm aren't exactly nonconformist these days. However, do not fear. The public chains will always be controlled by the public. No corporation controls these chains. The participating nodes control them. I have to say, honestly, that we still don't know how all of this is going to play out. But you can be sure that the corporate and government takeover of blockchain will be focused on private chains (and the financial doings of public chains). But in my not-so-humble opinion, the broad existence of blockchain technology will naturally outwit any censorship or attempts to bring it fully under anyone's umbrella.

Many people believe that the blockchain partnership supply chain has much more potential to surface creativity than does the traditional hierarchical supply chain. What's more, a great potential benefit of blockchain is making secrecy less desirable. Embracing this requires a leap of faith, but many now believe that transparency will help businesses thrive in the coming decades.

The concept of collaboration will become a heretofore unrealized reality, as businesses become comfortable sourcing ideas and solutions from far-off sources that they might never have come into contact with. Disparate loners or groups of people will more easily share ideas, work together, and come up with answers to questions that might not even have been asked. As part of this process, the centralized company will necessarily "give up" power. This is the hard part. Yet in giving up centralized power, the company will gain unprecedented decentralized and distributed knowledge.

There are lots of blockchain fanatics (BF) coast to coast, but the New York City ecosystem attracts a particularly wide variety of preening geeks, from social justice warriors to dentists juggling cryptocurrency between implants. One evening, as I scrolled through my list of blockchain meetups, I began to wonder about my own status as a BF. I felt a little obsessed with the tech, which surprised me. I wasn't the obsessive type. I decided to take a walk to clear my head.

As I headed from my home in Chelsea toward Madison Square Park, I wondered why neighborhoods in Manhattan were still demarcated. Why was Chelsea, gentrified as it is, so different in feel from the Flatiron, also gentrified? For all the talk about the sad homogenization of NYC, in the Flatiron I still passed piles of garbage bags, graffiti, and eccentric people doing their thing on the sidewalks. Plus, everywhere smelled like weed. Does everyone in this city smoke? What if each neighborhood were a block on a giant blockchain? I wondered. What if all the data—people—were registered to their respective blocks, and as you walked through the city, each neighborhood referred to the previous neighborhood? What if all this knowledge were stored in the brains of each of the people wandering the city, and those who weren't nodes in this system were rejected as false actors and kicked right out of town? In this city, right now, weed would be the coin of the chain.

Wow, I realized, stepping into the park. Blockchain has burrowed completely into my head. I'm lost. I'm a fanatic.

Another day, I read in the *Times* about the artist Kevin Abosch, who claims to be creating blockchain tokens for all the streets in Manhattan. Collectors can buy the tokens for a few dollars each. Such a deal! Since the artist has released 10,000 of the street tokens, it turns out to be a pretty good deal for him, too.

The Trust OS

Trust starts with truth and ends with truth.

—*Santosh Kalwar*

This is the perfect technology for my long-lost teenaged self, in that it makes clear that our current trust in authority is misplaced. Why do I trust Chase Bank? Everything from the way it charges for services to the way it designs its web interfaces tells me the bank's operating completely in self-interest, and that I am a mere cog assisting its success. Yet, I stick with Chase, year after year, because the alternatives seem like clones, just with different-colored logos.

Because blockchain records are immutable, there is no need for an "authority" like Chase Bank to affirm transactions. Whenever you make a transaction, it is recorded on the blockchain so that all who have access to that chain can see it. You can remain anonymous, but your transaction cannot. Thus, blockchain serves as a robotic generator of unconditional trust between humans. Unconditional, because no trust is necessary. Everything is guaranteed by algorithm. With guaranteed trust, there is no fear of fraud. That is a freedom that we rarely experience in the modern world.

If you spend money, such as bitcoin, litecoin, or other coins, the transaction is recorded. If you sell a horse to your neighbor, it is recorded, and the new ownership is clearly established. That neighbor, in turn, can easily sell fractions of the value of that horse to another neighbor. You can prove that by looking at the record, without having to appeal to (and pay) higher authorities to confirm the information.

This simple fractionalization of ownership, based on the mechanical trust built into blockchain transactions, could be a huge economic game changer. Currently, to divide an apartment building into cooperative shares requires lawyers, brokers, bankers, and others, takes ages, and costs a fortune. Smart contracts and cryptocurrency on the blockchain could make the process of fractionalized ownerships, and the trading of those shares—even the tiniest fraction of shares—straightforward. A renter in a building could purchase twenty-five dollars a month worth of coins that represent value in that building, and slowly, over time, build meaningful ownership. Not only would that be financially rewarding to the renter, but it might also increase the renter's sense of pride in the building, which might influence how she takes care of her apartment and the public spaces.

As anyone who has ever worked with an accountant, made a business plan, or kept track of calories day to day knows, the truth, as we express it, is malleable. In a sense, we all manipulate the truth to suit our needs and desires. Blockchain challenges that way of thinking—and thus the foundations of our entire society—as it stores evidence of transactions, creations, and historical occurrences. Let's say I work in the diamond business, and I send an email to my boss, blowing the whistle on a supplier who is hiding the fact that he's supplying us with blood diamonds. That letter will be forever represented in a block on the chain. No one will be able to deny that the email was entered as "evidence" at that time. Unlike a normal email, its registration on the blockchain couldn't be falsified.

Likewise, if we've put our supply chain on the blockchain, we'll know each step of the way where that diamond has been. With previous systems, the various players on a diamond supply chain—mining, transportation, cutting, setting, sales—often had separate ledgers that were hard to reconcile and easily subject to fraud. With blockchain the information entered cannot be altered and can be seen by all participants (some chains are public, others are private enterprise chains within organizations).

This is already happening. Three companies—Fura, Everledger, and DeBeers—are setting up blockchains to track diamonds from mining to sale, as a way to prevent trafficking in blood diamonds. The diamonds are certified conflict free at the source, and that certification is entered on the blockchain and will follow each individual diamond up the supply chain.

In distributed blockchain systems, every participant plays a role in determining the validity of transactions and information. Every participant contributes to the security of the system. A Mumbai peanut vendor using a five-dollar smartphone can keep a digital wallet on the blockchain and vote on the structure of smart contracts along with a private school teacher in Manhattan. The individual nodes might elect other nodes to represent them in making decisions, or they might vote, digitally, on various propositions before them, such as what dividends to pay or investments to make. With blockchain, poor people gain the power to make financial transactions, without having to be approved by banks and other centralized authorities. No one is shut out of the marketplace, and the people at the bottom of the supply chain understand better what is happening at the top.

Capitalism finds a new path. Melanie Swan says we'll all be crypto-citizens using blockchain to run our worlds. We will turn to ourselves, and the other nodes on blockchains, for the answers, rather than to the gatekeepers that now dominate our lives.

We each have our own definition of trust. As mentioned, many people use the term "trustless" to describe blockchain. Trust and trustless are two sides of the same coin. Most of us probably don't have a definition for trustless. At first, it sounds sinister, as in untrustworthy. However, in terms of blockchain, trustless is the trustworthiest possible state. Trustless means no trust is necessary. It's built in.

For centuries, we have used double-entry bookkeeping, a system that probably began with Arab scholars in Egypt, but was also used by Jewish merchants in medieval times, and by Korean mathematicians. Yet it wasn't widely used in Europe until the late fifteenth century.

In 1458, Benedikt Kotruljević, of Dubrovnik (then known as Ragusa), published a book describing double-entry accounting. He proposed that when a transaction occurs, there is a debit and a credit, and that these two columns must always match, or else there is an error. The ledgers of both the buyer and the seller of an object, for instance, will reflect this transaction—and any discrepancies in the transfer of money. Ideally, this system helps establish trust, and also leads to a more complete understanding of every transaction, for each party involved.

In 1494, a partner of Leonardo da Vinci, a Franciscan friar named Luca Pacioli, published a detailed guide to double-entry accounting that has led many to call him the father of accounting. The House of Medici and other merchant families embraced the ledger system and used it to create a pathway of lending, with trust, that led to great wealth. Over the subsequent 500-plus years, double-entry bookkeeping became the foundation of all the world's economies.

There's a cost for this trust—the fees of middle agents (in the tradition of the Medici) who certify transactions, including brokers, bankers, and others. In the United States alone, the top four accounting firms together earned $55 billion in revenue in 2017. Right now, countless accountants and bookkeepers sit at their computers attempting to reconcile their ledger reports with bank reports, sales reports, and the reports of other companies in an effort to ensure that everything is right in the world. That's because trust stops at their ledgers. They have no certainty about the trustworthiness of their counterparts' ledgers.

Ledger A		Ledger B	
Debit	Credit	Debit	Credit
7			7
	3	3	
	4	4	
8			8

DOUBLE-ENTRY ACCOUNTING LEDGER

Witness the financial crisis of 2008, in which firms such as Lehman Brothers were found to have used deceptive accounting practices that hid debts and overstated assets. Look at Enron. Look at Madoff. In effect, these firms had two sets of books, or more, and used cooked books to present themselves in a favorable light. This is possible because traditional double-entry accounting does not allow for immutable entry of assets and transactions.

Blockchain allows for more nuanced accounting that is permanent and verifiable. A firm could record transactions on a chain that included subledgers, such as how much the firm was owed on related transactions, what it owed others, and other useful information for determining the viability of the current transaction. Each transaction would begin with a unique contract, registered on a chain and visible to all parties involved, that would track any issues that came up during the execution of that contract. This would allow all parties involved to see the health of any transaction clearly and track all money exchanged. In effect, blockchain allows for adding a third column, verification, to the old double-entry accounting system.

For simplicity's sake, the blockchain world calls this triple-entry accounting.* Debit/Credit/Verification. Every transaction is agreed to by all parties at the time it occurs. There is no doubt, no confusion, no cheating. No Enron. No Madoff. And this trustlessness (meaning no trust necessary) is inexpensive — just the cost of transactions on the blockchain, without fees for banks, agents, and insurance.

---

* I'd like to offer a shout-out to Yuji Ijiri, the Japanese-born accountant and professor at Stanford and Carnegie Mellon who, largely unknown outside of accounting circles, in 1998, published a complex theory of triple-entry accounting, titled "A Framework for Triple-Entry Bookkeeping." Much of his theory is now being explored for accounting on the blockchain, but the current use of triple-entry is not entirely consistent with the ideas of the man who introduced the term. Ijiri died in 2017, obscure to everyone who isn't deeply involved in accounting. I can find no public comment from him about blockchain or cryptocurrency.

Decentralized, triple-entry accounting allows many new ways of doing business. Currently, the gatekeepers that control the flow of money and information also inhibit the flow of ideas. I don't believe this is necessarily intentional. Rather, it's a by-product of control. In the future, when information can be exchanged more easily between nodes, the formerly bankless farmer in, say, rural Amazonia might be able to share a strong idea for marketing products with the distributor in Manaus. Transactions—including the exchange of ideas—that might have been impossible through a central institution are done more easily person to person, thanks to distributed blockchain networks.

Ledger A		Ledger B		Ledger C	
Debit	Credit	Debit	Credit	Debit	Credit
7			7	-7	7
	3	3		3	-3
	4	4		4	-4
8			8	-8	8

TRIPLE-ENTRY ACCOUNTING LEDGER

THE TRUST OS • 87

Okay, let's be frank: Despite my pro-blockchain leanings, I know there's a big problem with trusting information on blockchain ledgers of all kinds. That is, what prevents someone from entering false information in the first place? Clearly, once information is registered on a chain, all subsequent transactions involving that information will be accurate, based on the original entry. But what if that original entry is phony?

The issue of *garbage in, garbage out* is still very much alive on the blockchain. Blockchain has no built-in tools to ensure that information on a chain was indeed accurate or truthful when it was first registered. For instance, a farmer could falsely claim that a batch of spinach is organic, when it isn't. That spinach could deceive buyers all along a supply chain, straight into your mouth, with no one knowing the difference. (However, if the spinach is tainted with pathogens, and you get sick, the specific batch could easily be tracked down, thanks to blockchain.)

In our current system, this highlights the need for independent outside agencies to certify the provenance of goods. Fortunately, most industries are now covered by reliable standards organizations. That spinach would need to be verified by a USDA-accredited agency. An organic cotton grower might be certified by one of two agencies, Organic Content Standards or Global Organic Textile Standard. Likewise, legal and other documents can be certified as true and accurate, if necessary. Many times, it is the reputation of the person or entity registering the documents on a chain that makes the information believable. Reputation remains key.

IBM has designed small, readable "DNA" pellets that can be coded with information and placed inside bundles of goods and "read" at each point along the delivery route, so that less expensive goods can't be substituted. A company called Dust has created "fingerprints" made from diamond dust that can permanently mark any item with a unique code. Other promising ideas are in the works.

I agree that *garbage in, garbage out* could be an issue for some blockchain services. However, I also believe that blockchain systems are designed to encourage honest behavior. That's because while data is permanently registered on the chain, modified data referring to the original entry can also be permanently entered on a chain. So, if someone's data proves to be false, down the road that will be noted on the chain, to the detriment of that transaction and also, potentially, to the reputation of whoever registered it. If that dishonest broker is found out, then the whole world will know, and their future transactions will not be trusted. Again, reputation remains key.

Once a transaction takes place on a blockchain, all future transactions of that product, coin, or intellectual property will reference all the previous transactions. This is why it's impossible to spend one digital bitcoin in two places, for instance. The system is designed to promote good behavior.

Trust takes on another dimension when humans interact with machines. Some of the ideas that blockchain enthusiasts propose induce outrage, disbelief, feelings of stupidity, and complete lack of comprehension in those who aren't (yet) enthusiasts.

Consider this: You and some friends buy a self-driving car and join it to a computer network that connects people in need of a ride to cars in need of a fare. Not Uber or Lyft, which link customers and drivers through their centralized website. You set the car up with smart contracts that let it run itself, with the idea that at some point it will purchase itself from you for a good price, and you won't have to think about it any further. In this network cars have a direct connection to all of the customers, and vice versa. When the car is summoned, it picks up the passenger and heads to their destination. When the ride ends, a string of code in the network automatically pays the car from the passenger's account. Then the car drives itself to an electric vehicle charging station, fills up on electrons, and another string of code pays for the service.

The car "happily" picks up passengers, 24/7, stopping now and then at a carwash to get its interior cleaned, until one day it gets a flat. An algorithm contacts a repair shop that sends a human or robot repairman out to fix the tire. Again, the string of code releases payment to the repair shop. Eventually, the car amasses a substantial amount of money in the bank account that was set up so it could manage itself. It uses the money to buy itself from you, in an offer that's too good to resist. The self-driving car continues working, saving almost all of its earnings (it has no need for lattes, plane tickets, health care, or fresh-cut flowers), and soon has enough to buy another autonomous vehicle.

That's right, the car buys a car. Then it lets that car purchase itself. Repeat. Repeat. Repeat. In time, this one driverless car has expanded the network by hundreds of other cars, all of them autonomous, all with equal participation in the workings of the network. Does this company inspire fear, or trust?

The cars, blockchain, and smart contracts aren't yet sophisticated enough to run a successful DAO, or distributed autonomous organization. While I have doubts about whether a DAO like this might one day succeed, I do strongly believe that distributed networks will change the way we do business going forward. They offer too many advantages to be ignored.

One distinct advantage of distributed and decentralized networks is that they are more resilient against attacks than other networks, for several reasons:

- Because they involve dozens, or hundreds, or even thousands of devices, these networks won't easily fail.

- They are harder to hack, for the same reason.

- They are resilient, because if one device, or node, fails, the others will pick up the slack. In fact, when data is shared across all nodes, every node or device on the network but one can be destroyed, and that one device will preserve the integrity of the data, at least for a time. All it takes is one MacBook or Dell to survive the attack.

III

MONEY AND CREATION

The Person Behind Blockchain

It's very attractive to the libertarian viewpoint
if we can explain it properly. I'm better
with code than with words, though.

—*Satoshi Nakamoto, November 14, 2008*

The worst thing about blockchain is that an engineer named it. It's the brainchild of a mysterious entity named Satoshi Nakamoto, who used blockchain as the base for bitcoin, the digital currency he invented and announced in 2008. At the time, few people appreciated the usefulness of blockchain technology. We can't ask him what he truly thought in terms of branding—probably not much, considering his equally awful choice of bitcoin as the name of the new currency. That's because he's been offline, out of communication, since 2011. If anyone had asked, I'd have named the system Satosh, and the coins Satoshis. But no one asked.

By the way, most people who bother to think about it quickly come to believe that bitcoin began with a deception. It is very likely that Satoshi Nakamoto is a pseudonym. For the purpose of clarity, I'll refer to Satoshi as "he," although we don't know his gender, or whether he was one person or a group of people working together. Curiously, Nakamoto's white paper, a short, authoritative report explaining the method and philosophy behind bitcoin, betrays absolutely no affinity for the Japanese language, or culture. Instead, some say the language betrays an upper-crust English education. There's no way to know the truth, at the moment, as Satoshi never revealed anything about his personal life or where he lived.

In November 2008, Satoshi posted his white paper about the bitcoin protocol on the Cryptography Mailing List. The post opened with these words:

Bitcoin P2P e-cash paper
Satoshi Nakamoto Sat, 01 Nov 2008 16:16:33-0700

I've been working on a new electronic cash system that's fully peer-to-peer, with no trusted third party.

The paper is available at: http://www.bitcoin.org/bitcoin.pdf

The main properties:

- Double-spending is prevented with a peer-to-peer network.
- No mint or other trusted parties.
- Participants can be anonymous.
- New coins are made from Hashcash style proof-of-work.

The proof-of-work for new coin generation also powers the network to prevent double-spending.

You can still read the short paper in its entirety online. At the time it was published, not many, outside of a small group of cryptocurrency fanatics, actually read it. These fanatics had plenty of reason to be excited. Ecash, the first digital cash, was created in 1983 by David Chaum, who also invented a payment system called Digicash, in 1995. The National Security Agency looked into the concept of anonymous digital cash that couldn't be traced. In 1998, Nick Szabo invented bit gold,* with similarities to bitcoin† that appealed to many people. These early forms of digital cash were not distributed or decentralized—unlike Nakamoto's invention, all of these currencies depended on the hierarchical structure of banks.

In 2009, soon after publishing his white paper, Nakamoto released the open source bitcoin software to the world for anyone to download and use for free.

In the early days of bitcoin, Nakamoto collaborated with others to improve the code, and geeks began to play with it. Nakamoto communicated by email with a few bitcoin developers in the following months, but no one has heard from him since 2011, when he sent an email to Mike Hearn and a few other developers. The group included Gavin Wood, whom he'd picked to run the Bitcoin Foundation, a nonprofit that nominally tries to govern the bitcoin blockchain. According to Hearn, "He told myself and Gavin that he had moved on to other things and that the project was in good hands."

A number of investigative journalists have tried to track down the

---

* This is different from BitGold, a cryptocurrency currently traded in crypto exchanges.

† Many people speculate that Satoshi Nakamoto is actually Nick Szabo. However, Szabo, who is currently a thought leader in cryptocurrency, denies it.

"real" Satoshi Nakamoto, yet no one has presented convincing evidence. For example, in 2014, *Newsweek* reporters followed a series of clues that led them to the front door of a sad suburban home in the San Gabriel Mountains outside LA. The owner, a man named Dorian Satoshi Nakamoto, called the cops (who wondered why the inventor of bitcoin would be living in such a down-market home). He also hired a lawyer and issued this statement: "I did not create, invent or otherwise work on Bitcoin." He's easy to believe.

Curiously, while the inventor of bitcoin is said to have mined 980,000 of the coins for himself, not one of those coins has ever been traded, according to the public records of the bitcoin blockchain. That's strange, considering how valuable they've become.

At first, the coins were worth zero dollars. As people joined the bitcoin forum, they began negotiating the price among themselves. In May 2010, a bitcoin owner going by the name of "SmokeTooMuch" tried to auction a lot of 10,000 bitcoins for $50. No one bid. Months later, bitcoin were valued at a fraction of a penny. That same month a crypto-explorer named Laszlo Hanyecz, based in London, posted the following request in Bitcoin Forum:

Pizza for bitcoins?
May 18, 2010, 12:35:20 AM

Merited by Seccour (50), alani123 (12), OgNasty (10), the_poet (10), leps (10), mnightwaffle (10), arthurbonora (10), cheefbuza (7), d5000 (5), Betwrong (5), mia_houston (5), klondike_bar (3), malevolent (1), EFS (1), vapourminer (1), iluvbitcoins (1), ETFbitcoin (1), HI-TEC99 (1), S3cco (1), jacktheking (1), LoyceV (1), bitart (1), batang_bitcoin (1), Astargath (1), coolcoinz (1), apoorylathey (1), Kda2018 (1), TheQuin (1), Financisto (1), Toxic2040 (1), amishmanish (1), Toughit (1), nullius (1), lonchafina (1), alia (1), inkling (1)

> I'll pay 10,000 bitcoins for a couple of pizzas.. like maybe 2 large ones so I have some left over for the next day. I like having left over pizza to nibble on later. You can make the pizza yourself and bring it to my house or order it for me from a delivery place, but what I'm aiming for is getting food delivered in exchange for bitcoins where I don't have to order or prepare it myself, kind of like ordering a "breakfast platter" at a hotel or something, they just bring you something to eat and you're happy!
>
> I like things like onions, peppers, sausage, mushrooms, tomatoes, pepperoni, etc., just standard stuff no weird fish topping or anything like that. I also like regular cheese pizzas which may be cheaper to prepare or otherwise acquire.
>
> If you're interested please let me know and we can work out a deal.
>
> Thanks,
>
> Laszlo
>
> BC: 157fRrqAKrDyGHr1Bx3yDxeMv8Rh45aUet[*]

He got his pizzas. That roughly $40 worth of bitcoin would have increased to nearly $200 million in December 2017, when the value of individual coins rose close to $20,000. (To illustrate the extreme volatility of bitcoin, those pizzas would have dropped in value to about $64 million by late August 2018, when bitcoin value dropped to about $6,400.) The rise in value created many bitcoin millionaires. Satoshi Nakamoto's unspent coins are easily worth $10 billion. Satoshi's stake is worth about ten times the total market cap for the entire Papa John's Pizza chain.

---

[*] Hanyecz tried to replicate this purchase in 2018, using the lightning network. This elaborate purchase showed just how convoluted commerce remains with bitcoin, as Hanyecz had to "subcontract" the work to a friend, paying this middleman to arrange the pizza delivery, rather than just sending the coin directly to the pizza shop. Kind of lame, although it does illustrate the value of coins now compared to his first pizza purchase.

2010—2 pizzas 10 000 BTC
2018—2 pizzas 0.00649 BTC

The Meaning of Money

Money often costs too much.

—*Ralph Waldo Emerson*

Bitcoin is an odd invention, and also a vivid illustration of how a nonhierarchical system, with no leader or marketer, is shaped by its users. Take the way people represent the coin on websites, tattoos, and T-shirts. Most illustrations of bitcoins show them to be shiny and gold, with a giant "B" in the shape of a dollar sign, or the Thai baht. Some are illuminated from within. Some spew rocket flames. Others wear gold and purple crowns.

These fantastical images evoke easy success. Not surprising, I think, since 71 percent of bitcoin owners are male, and 58 percent are young, according to a poll of nearly 6,000 owners conducted jointly by Survey Monkey and the Global Blockchain Business Council. It seems natural that these images appeal to the same testosterone-driven sensibilities that keep superhero comic books flying off racks. Almost every coin meetup or conference I've been to has been filled with at least 80 percent men. The online forums are overrun by macho-nerd smashups who are inspired to take the bull by the horns, get rich, and grow a pair, for a change. Bitcoin turns nerds into swaggering bros.

The iconography of bitcoin appeals to flashy male fantasies meant to make the coins seem more real. You could fall for a coin like that, touch it, show it off. If you are smart you'll capture those golden flames yourself and get the bitcoin effect: money, wealth, riches. It's more than greed that drives the bitcoin craze, it's a desire for power.

If only I'd bought a few thousand coins back in 2010, the thinking goes, I'd be a winner right now. A king with enough money to run for president. Currently, men hold about 95 percent of total bitcoin value. It's a glaring gender division.

Yet anecdotal evidence and my own two eyes tell me that women are drawn in great numbers to the other uses for blockchain, including decentralized companies, distributed networks, and other world-changing manifestations. Coins are brawny, but the networks are going to be truly powerful. The real superhero nature of blockchain will be evident when the utility of the coins is joined with the paradigm-shifting structures of distributed systems to create new businesses, communication styles, and governments that all of us can embrace. The coins have the potential to be far more than investments. They are poised to unleash blockchain potential in ways we can't yet imagine.

Gold vs. Code. There are thousands of cryptocurrencies, such as bitcoin, ether, and litecoin, at large in the world. There are also probably hundreds of millions, if not a few billion, naysayers, critics, and pooh-poohers, including the Oracle of Omaha, Warren Buffett.

In early 2018, Marketwatch quoted Buffett as saying, "You can't value bitcoin, because it's not a value-producing asset." A few years earlier he had said, "Stay away from it. It's a mirage, basically. The idea that it has some huge intrinsic value is just a joke, in my view." Buffett has also admitted he doesn't understand how the technology works. That would seem to undercut his argument.

The most common complaint about these currencies, from Buffett and others, is that they aren't backed by anything. There are no precious metals guaranteeing the value of bitcoin, not even a warehouse full of gold-plated vaping pipes, or beard oil, or anything else of value. Still, the curious thing is that these same people often wouldn't hesitate to invest in gold, if the price were right. And there is nothing guaranteeing the value of gold, other than desire and the market, and some pretty rocks.

Gold and bitcoin have a couple of things in common. They are both, essentially, commodities. Yet both are also considered currencies, because of how they are used. Desire, trust, and the marketplace give both gold and bitcoin value. In those regards, they are no different from the U.S. dollar.

Bitcoin haters often refer to the U.S. dollar as the standard to which all currencies should be compared. That's curious, when you think about it. According to the Treasury Department, the dollar is guaranteed by "the full faith and credit of the U.S. government." While that is a reassuring statement, it is not a tangible asset.

This is from the Federal Reserve's website:

Is U.S. currency still backed by gold?

Federal Reserve notes are not redeemable in gold, silver, or any other commodity. Federal Reserve notes have not been redeemable in gold since January 30, 1934, when the Congress amended Section 16 of the Federal Reserve Act to read: "The said [Federal Reserve] notes shall be obligations of the United States. . . . They shall be redeemed **in lawful money** on demand at the Treasury Department of the United States, in the city of Washington, District of Columbia, or at any Federal Reserve bank." Federal Reserve notes have not been redeemable in silver since the 1960s.

The Congress has specified that Federal Reserve Banks must hold collateral equal in value to the Federal Reserve notes that the Federal Reserve Bank puts in to circulation. This collateral is chiefly held in the form of U.S. Treasury, federal agency, and government-sponsored enterprise securities.

In other words, the U.S. dollar is backed by debt, along with faith in the U.S. government. Bitcoin is backed in part by the scarcity that results from the limited supply of algorithmically generated coins; the mathematical "proofs" people must solve in order to earn new coins; and the game-theory implications of its design that compel people to make the network secure. There is no need for trust or faith in bitcoin, as long as you have faith in the algorithm. Many people don't. Some don't because they misunderstand how it works. Others say they understand it all too well, believing it to be some sort of pyramid scheme, or tulip bubble. The concept of creating currency from thin air can be difficult to register, although that's essentially how our government prints the U.S. dollar.

I think both currencies are quite strong, and likely to be around for a long time. That said, people should follow their instincts and intellects when investing in cryptocurrencies. Even Vitalik Buterin, who cofounded the Ethereum blockchain and its currency, ether, agrees.

In a tweet he wrote that "people who do not hold cryptotokens are not uncool losers. They are people using very reasonable heuristics in not personally participating in an industry that they do not understand and there are many justifiable reasons to be wary of."

Three items of value gather under the modern cryptocurrency umbrella: coins, altcoins, and tokens.

While there is no set definition, in general, bitcoin is the original and only coin. All the other similar currencies, such as litecoin, ZCash, Dash, Namecoin, and more than 1,500 others are considered altcoins. Most people view them as currency, because they have a value and can be exchanged. In some cases, you can use them to buy things, though cryptocurrency isn't widely accepted on the retail level.

All of these coins represent abstract value, and they generally are issued by algorithms. It's believed they can't be easily manipulated, though they haven't been around long enough to prove this out.

Tokens can be largely indistinguishable from bitcoin and altcoins, which I'll just refer to from now on in the aggregate as coins. But they have the additional value of being able to represent goods, such as clothing, water from a well, or even a share in an apartment building. They generally do not reside directly on the blockchain, but are linked to it from software "layers" above the chain. They can do a lot more than coins, and faster.

Tokens and coins, when combined with smart contracts, can pay themselves out when, for instance, the terms of a contract are satisfied. The tokens and coins can also be divided into minuscule fractions, making it possible for people to purchase shares of expensive items. For instance, it's conceivable that a person could buy a $1,000 share of a $10 million de Kooning painting and trade that share with ease on the blockchain. This proposition is already being tested in real estate, on a small scale, by a Brooklyn company called Meridio.

While all sorts of regulations stand in the way of this at the moment, there's a lot of hope that things will change. These coins and tokens could truly democratize investing, leveling the playing field so that the poorest among us could have ownership, even if it's a small fraction. The potential changes to society include increased pride of ownership, behavior changes, increased wealth, and less significant class divides.

Programmable money is either an exciting development or a sign of the apocalypse, depending on where you are coming from. But whatever you think of the stuff, it's here to stay. Programmable money is cryptocurrency combined with smart contracts that can represent any kind of value. The coin can, in effect, trade itself according to algorithmic instructions embedded in the contract. In this way, coin-embedded software will make economic decisions for us. You just give it "If X happens, then you do Y" instructions. For instance, if Party 1 signs the correct documents, then Party 2 releases money to pay for the house, with the transaction and deed being recorded on the blockchain. It doesn't matter if the two parties trust each other. That's made irrelevant by the certainty and durability of the smart contract. These programmable coins would be tied to specific values, products, and services, rather than being spread across the value landscape like a common dollar.

An example of how tokens can factor into business is the use of blockchain ledgers to revamp and improve loyalty rewards programs at supermarkets, airlines, and stores, replacing points with universal blockchain-based tokens that might be more effective for customers and businesses. What if hundreds or thousands of different businesses gave out a uniform loyalty token, which could be redeemed at any participating business, or traded among customers in an unregulated aftermarket?

The cost of an aggregated token system like this would be far less than the sum of all the individual rewards programs. And it would minimize the problem some businesses have when too many people at once redeem their rewards points, thus hurting cash flow.

For customers, universal tokens would be far more useful than points tied to just one company. You could, for instance, use tokens earned from an airline to pay for a massage. While some loyalty programs currently cross-pollinate, there is always the question of how value is determined when points from one company are used for another company. Universal tokens solve that problem. They wouldn't expire, and if the initial company you earned the rewards from went out of business, your tokens would still be valid elsewhere. BitRewards and Blockpoint are two companies working within this space.

What does money mean when robots earn it and spend it, without human intervention? For my money, *The DAO* is the most potentially transformative organization to have come out of the blockchain revolution. It's also one that in its weirdness is difficult to grasp. Yet diving into the story and philosophy of this distributed autonomous organization cracked open my imagination, leaving me wondering what effects it would have on the future of business and organizations.

The DAO had nothing to do with the ancient Chinese philosophy of Daoism, although I do see similarities. One interpretation of Daoism is that it is the unfolding of reality, as we know it, at the same time that this very reality is transforming itself into something new. That is to say, it is the evolution into the present at the same time that it is the transformation of the present into the future.

Parsing *The DAO* is about as easy as parsing the Dao. Yet in its brief flaming trajectory, the very nature of *The DAO* was transformational, affecting all consciousness that followed.

Writing in *Tech Crunch* in May 2016, Seth Bannon said of *The DAO*, "A new paradigm of economic cooperation is under way—a digital democratization of business." That's a curious way of describing a business that transformed itself into a failed experiment—but it is true. *The DAO* has inspired many business thinkers ever since.

The DAO was a DAO, of course, like the self-owning cars I've already described. However, *The DAO*, founded in 2016, was the first company of its kind. Designed and set loose primarily by early participants in the Ethereum blockchain, it operated according to a mysterious new system that supposedly had no leaders, and no center. This *DAO* was composed of nodes of investors that could vote and transfer value using smart contracts and other dapps. It was designed to be a venture fund working primarily with cryptocurrencies and decentralized organizations. Proposals to the fund would then be voted on by members of the chain.

The DAO was created and guided by humans, who encoded the rules into smart contracts. But future DAOs might be operated automatically, run by machines with little or no human input. Eventually, AI and other technologies will make it possible for machines to create smart contracts for other machines, with no need for human intelligence. Algorithmically driven smart contracts will manage every transaction and creation, working together to accomplish tasks and make money or meet other goals.

While The DAO was a group of people who acted as a single economic entity, it had no official recognition as a corporation and was not bound to any state or government. All the members were given tokens that allowed them to vote on governance and which new projects to fund.

Anyone who wished could submit a business idea to The DAO, and the members would vote on whether to fund it. Decisions were based on the ideas and smart contracts proposed, rather than on the people proposing them. This leveled the playing field, as traditionally, entrepreneurs with connections and cash have been more likely to get funding for their enterprises. Anyone with a good idea and good code for a business had the chance to be funded. This is an important part of the paradigm shift.

The DAO itself was crowdsourced, netting more than $150 million in ether coins. As ether rose in value, that amount rose with it. At one point The DAO was "worth" $250 million. Anyone who invested in the enterprise was considered to be an owner. Still, no sooner had it ballooned in size than hackers discovered a bug that let them continually extract more ether coins from The DAO than they were allowed

THE MEANING OF MONEY • 113

to. *The DAO* noticed it, in the way that things get noticed quickly on blockchains, and began correcting the leak. Yet it wasn't plugged until tens of millions of dollars' worth of ether had been stolen.

Unsurprisingly, this upset the investors in *The DAO*, many of whom were early converts to the Ethereum blockchain ecosystem on which *The DAO* was built. They wanted their money back, and they knew a way to get it. Ninety-eight percent of the nodes agreed on a plan to take advantage of the concept of the dreaded 51 percent attack, and go back in time and rewrite the code to make the stolen coins obsolete and restore the value to the "rightful" owners. This would involve making a "hard fork," or new path for the blockchain to follow, leaving the old path alone, along with the potentially worthless coins that had been stolen. Then they could restore the funds that had been tampered with by creating new ether. This was anathema to the hard-core blockchain believers, who respected the chain as an immutable series of transactions that should never be overridden in this way by a "centralized" authority.

These dissidents felt that, in effect, a ruling class had been so devastated by the theft that they arbitrarily forked the chain, without getting consensus from all the nodes. They even accused this group of rigging the vote so they'd get their coins back. The critics called it a bailout (hard to argue with that) that went against the blockchain ethos, which said the chain, including any DAO, should run itself. One critic said the Ethereum chain would forever after be known as "the chain of thieves."

If the "ruling class" indeed had cheated, they'd done a very good job, as the hard fork, which took place on July 20, 2016, will forever be memorialized in the Ethereum chain's 1,920,000th block. You can check the block out yourself, by googling "Ethereum block 1920000." The Ethereum chain, of course, thrives. *The DAO*, however, soon was abandoned.

The DAO **was meant to serve as a flagship example of how distributed systems could function apart from the normal rules of business.** As I've noted, part of blockchain's security comes from the fact that it would take 51 percent of the miners banding together to challenge the content of blocks on the chain—called a 51 percent takeover. This is deemed largely impossible, as chains get larger, with more nodes involved in certifying transactions. Still, in the case of *The DAO*, the owners of the lost coins banded together with a large enough percentage of nodes on the chain and just declared their old coins obsolete, while minting new ones. Many purists felt the chain should have proceeded without interference. They saw the coin play as corrupt.

And it was. A group of self-interested people were able to subvert the idealistically modeled system they had set up as a beacon of distributed thinking. It was a grand experiment, and it failed. The coding was faulty, and somebody stole a lot of money. Rather than see where *The DAO* went after that, these people put their money in front of their ideals and rewrote the code to get it back. Given that we're talking about millions of dollars, their behavior was totally predictable, yet tawdry.

More than 1,600 altcoins have been created since the birth of bit-coin. Since 2016, most of these have fueled Initial Coin Offerings, a highly controversial method of raising money to fund a company—or to abscond with when you drop out of society for good. Here's a list of a few coins you can buy:

PotCoin

ZCash

TitCoin

Cardano

Stellar

In an ICO, a person or group of people with an idea for a business create a coin and sell that coin to deep-pocketed people in order to raise money. Technically, these aren't investors, because they aren't investing money in the company, but rather are buying coins issued by the company. They purchase the coins based on the idea behind the new company, the team that is going to run it, and a white paper that outlines in great detail just what is going to unfold. If they think the coin will rise in value as the company evolves, or that a separate market for the coin will develop, they will invest. The investors generally pay for the new altcoin with established coins, such as bitcoin and ether, that they have purchased with dollars or other state currency. In 2017 and 2018, billions of dollars were raised in ICOs in the United States, much of it by reputable companies, such as Codex Protocol, a company that provides proof of provenance for artwork and collectibles sold by auction houses.

There were also some big scams. Because the SEC has been slow to regulate cryptocurrencies, until recently, these ICOs have not been subject to intense government scrutiny. In many cases, unscrupulous (and code-smart) players have set up ICOs and taken the money and disappeared, knowing that nothing required them to build a company and make money for everyone. ICOs are risky enterprises to invest in. A study by an ICO advisory firm named Statis found that 80 percent of 2017 ICOs were scams. That is an astounding number. At the same time, in the end only 11 percent of money spent on ICO coins went to these scams. Apparently, most people can spot a good deal, or a bad deal, when they see it.

Savedroid was one such bad deal. It claimed it was going to use AI to manage an investment fund and also offer a crypto-credit card. Dr. Yassin Hanker, the founder, raised $50 million and then just shut the operation down. The people who bought the now valueless coins had no recourse. Hanker posted a tweet saying, "Thanks guys, over and out," along with two photos: a selfie and a hand holding a bottle of beer on a beach.

Following the mania of 2017, people looking to buy in to ICOs seemed to begin vetting the white papers and teams more seriously. And the SEC and other government agencies started paying more attention. The ICO model, which puts the horse before the cart by funding a company before it has a viable product, could transform how entrepreneurs succeed in the future. ICOs are a powerful tool for creative builders who don't have access to traditional finance, or venture capitalists. These fund raises can be color-blind, status-blind, and gender-blind ways for previously unfunded entrepreneurs to bring their ideas to life. We just need to weed out the bad apples and charge forward.

An interesting use of a very small-scale ICO for small business was until recently on display at www.evancoin.com. This was where you could buy evancoins to be used to purchase an hour of fuzzy.ai cofounder Evan Prodromou's time. Or you could hang on to the coins and redeem them later or trade them with others. Prodromou accepted coins in return for business consulting time only—he didn't charge his family, for instance, to hang out with him on a weekend. Prodromou's ICO, in October 2017, was quite modest: He sold twenty coins at $15 U.S. each. Within two weeks, the value had risen to $45 per coin. After that, he released many more coins. Now the site appears to be compromised, and anyone visiting it is given a warning that it is not secure.

That highlights some of the security problems with establishing coins and building dapps on top of blockchain. Too bad, in this case, because before it became marked with a security warning, the evancoin website offered an illuminating description of how to spend one of the coins:

How to Spend EvanCoin

1. **Get some EvanCoin.** You can buy some EvanCoin, or you might be able to get some for free.

2. **Send me 0.01 EvanCoin.** (optional) This is definitely optional, but it tells me that you know how to transfer EvanCoin to me, and it will cover the 30 seconds it takes for me to read your email. My Ethereum address is 0x001be02a4742767000cc54a820686a3087e4d472. You should be able to transfer tokens in your wallet, unless your

wallet is MetaMask,[*] in which case follow the instructions
for Managing tokens with MetaMask.

3. **Send me an email telling me what you want to use
my time for.** I'm not going to do things that are illegal,
humiliating, or abusive to others. If you want to check for
some suggested usage, and rough costs, please look at how
to use Evan's time.

4. **Send me the EvanCoin.** Send me the total amount,
where 1 coin = 1 hour (or fraction thereof), and I will do
the thing you ask. If you aren't satisfied, or if I can't do it,
I'll refund part or all of your EvanCoin back. You can then
try to use it for something else, or sell it.

Admittedly, before it closed, EvanCoin appeared to be more of a
performance about how cryptocurrency and blockchain could work
than a viable business plan. But as such, it offered a creative vision that
I think might still lead to interesting individual ICOs in the future.
For instance, someone might fund their PhD in quantum mechanics
by selling coins now that could be redeemed after the PhD, when
presumably the person's time would be much more valuable. Or a
promising artist might sell coins now that people would buy with the
hope that the value would rise. In effect, the ICO allows individuals
to market themselves as solo investments.

---

[*] Metamask is a dapp that allows a person to access various aspects of the ethereum blockchain, using a traditional browser.

IV
THE THRILL
OF THE NEW

Stacks of Apps Called Dapps

Whereas most technologies tend to automate workers on the periphery doing menial tasks, blockchains automate away the center. Instead of putting the taxi driver out of a job, blockchain puts Uber out of a job and lets the taxi drivers work with the customer directly.

—*Vitalik Buterin, cofounder of the Ethereum blockchain*

Dapps, or distributed applications, open the blockchain to all kinds of purposes. They are essentially software that's built on top of the chain in another, more easily accessed layer. Dapps can be customized to accomplish many tasks, from paying someone for generating solar electricity, and charging their electric car for consuming it, to tracking the path your blackberries took from the farm in Mexico to the processor in California to the artisanal ice cream factory in Brooklyn. You would know if the berries were organic, as advertised, and whether they'd been washed. There are now dapps that let Estonians vote, that track the journey of alpaca wool from farm to sweater, and that let citizens make micropayments to journalists whose work they read (cutting out the influence of advertisers, owners, and editors). AID:Tech, a blockchain-centered women's health project in Tanzania, uses a dapp to follow the medical progress and care of pregnant women; in July 2018 the very first babies were born on the blockchain.

The business of health care is sorely in need of technological improvement. Blockchain is rapidly proving itself useful in this field, with a host of startups, including Gem, MedRec, and BitMED. There are three main areas where blockchain can help: medical records, consent to share those records, and micropayments to patients as a reward for healthy behavior that reduces overall medical costs. For instance, medical records can be written into blockchains, so that they can never be changed. The patient could control who has access, and to what, at any time. When used with clinical trials, blockchain would remove any temptation to manipulate the data. And patients might be rewarded for letting researchers use their data, as well as for sticking to their treatment protocol. BitMED is taking this a step further, by offering people free medical care in exchange for the right to sell their data to drug and medical device companies, and to researchers. According to BitMED, the global health data market is worth $230 billion. They say the data that is now taken for free would subsidize or pay for a person's health care.

Of course, no one debates the necessity of digital cats. They're meow as hell, cute as a scratch on the neck, and they don't carry toxoplasmosis: Clearly, CryptoKitties are the new cats. You can buy these little digital cat images that live in your phone with ether coins, using Ethereum blockchain technology. Like so many other figments of our digital imagination, CryptoKitties are DTF: You can breed them to create other, possibly more valuable kitties. The CryptoKitties company, in Vancouver, says these cats are collectibles, like baseball cards and art. Seems to me that, however you want to look at it, in the end the kitties are money.

You can't pet them, other than by stroking them on your screen. You can't take them to the vet, or wear them around your neck like a stole. They are graphic, hypoallergenic, and pretty obedient. Their value is determined by scarcity and desire, just like most things in our world. Apparently, desire is pretty high, because when the kitties first came out so many people bought the cats that the transactions slowed the massive Ethereum blockchain to a crawl. CryptoKitties has generated over $20 million in sales, with some cats going for over $100,000, according to the *New York Times*.

The population began with the release of "hundreds" of kitties. For the next eleven months, a new Generation 0 kitty was released every fifteen minutes or so. These first kitties don't have parents. But as they are bred, their offspring take on the previous generation's genetic qualities. You can pay other "owners" tens of thousands of dollars to breed their cats with yours, if you think their DNA is worth the cost. The holy grail is to breed a "Fancy Cat," by creating a special "ascension" kitty. Just keep trying until you get the right one.

The CryptoKitties company pockets the price of each Generation 0 cat that is released. Plus a 3.75 percent fee on each transaction within their marketplace. Meow! This blockchain thing might just have (four) legs.

Like hip-hop, ultimate Frisbee, and LSD before it, blockchain is clearly a culture as much as a thing.

I walked down New York's Canal Street late one afternoon looking for the cryptocurrency bohemians. Horns blared and the sidewalk was packed with grandmothers pushing carts full of Chinatown groceries home through the crowd. I was surprised at how little had changed in the thirty years since I'd first wandered this street. More tourists, maybe, shopping for off-market luxury goods sold by the African traders, oblivious to the fact that they could get hauled off to the slammer if the fashion task force were to take exception to their Louis Vuitton bag. I found my way to the entrance of a nineteenth-century warehouse and climbed an old-school flight of steep wooden stairs, opened a door, and was greeted by a young intern/receptionist. The loft was about eighty feet long and twenty-five feet wide, with worn plank floors and a few nonsensical glass "rooms." The ceiling fixtures were copper-colored Tom Dixon lamps that were popular in their day, though they looked like sad plastic now. It felt like a rich person's loft circa 2005.

"I didn't get a ticket in advance, so I'd like to pay here," I told the receptionist, who stood sentinel with a large iPhone. She typed and typed and typed into the screen and finally said, "Fifty dollars."

"Fifty? The website says thirty-five."

"Now it's fifty because of the crowd," she said. "Demand."

Congestion pricing, surge pricing, rip-off pricing, I thought. But I paid and took a seat. A screen near a makeshift bar up front displayed van Gogh's *Starry Night*.

Two thirtysomething women sat next to me. They were art curators for an all-female nonprofit gallery in Dumbo called A.I.R. They knew a bit about blockchain, but didn't seem obsessed. They were there to learn, but, looking around at the pretty dull crowd, and the cheesy van Gogh screen on the wall, they didn't get their hopes up.

Finally, the leader of this high-priced "meetup" took the floor and introduced the first speaker. This was the guy with the van Gogh. He had absolutely nothing to do with blockchain, but he did have this art biz called Meural, which manufactured and sold framed screens designed to hold digital art collections. You'd buy one and then subscribe to his service for forty bucks a year, which would give you access to an imaginary space where "the iconic meets the unexpected." You could download whichever paintings you wanted from their licensed collection of digital images. They claimed the collection represented $3 billion worth of art. But not really, since all the art was in pixels and you could not smell the paint. The speaker hoped someday to use cryptocurrency to reward artists each time their work was viewed or resold, but he didn't seem to have any specific plans. Applause. Behind me a young guy said to his curly-headed nerdy neighbor, "Why are you here?"

"I just bought my first crypto two months ago. I'm very interested in art on the blockchain."

Two months ago, I thought, laughing to myself at the absurdity of the speed at which this technology had rolled through culture. Jeez. I bought my first crytpo seventeen months ago. I was an old-timer.

One of my curator neighbors suggested I look at Rhizome, an online publication with archives of digital art. She said that's where digital art was happening.

Then things got really interesting. A lanky, clean-cut guy named Tommy Nicholas, founder of Alloy and special adviser to Rare Art Labs, spoke about using tokens, rather than coins, as a way of valuing art. His main point, and it's an important one, was that tokens only work if you create a community that values them. When they are valued, they become a sort of currency for rare digital art. My head almost popped off when I heard that term. What the hell did it mean?

Tommy explained: The first copy of a digital image, when it's registered on a blockchain, becomes an immutable work of art. There can only be one. All the screen grabs and other copies that are made and passed around will never have the provenance, guaranteed on the chain, of being the first. Thus, a rare digital image. Once this was an oxymoron. Now it isn't.

Next up was Matt Hall, one of the creators of cryptopunks, collectible digital rare art of cheesy-looking pixelated punk rockers. They released 10,000 of these characters into the marketplace, with tokenized values, and watched them trade. People have made hundreds of thousands of dollars off these babies, all tracked and traded via blockchain. Now that's rare digital art!

As the evening moved slowly along into a snooze, my new curator friends ordered takeout on their phones (using old-fashioned credit cards), and my mind kept going back to cryptopunks. They seemed so of the moment, and also so nostalgic (punk rockers?), so innovative as an art product and yet so dull.

Leaving, I felt full of blockchain thoughts, and happy once again. The steep wooden stairway to the street gave me nostalgic thoughts of NYC in the late seventies, when these lofts went begging for tenants, and you never knew what you'd find at the top of the stairs: a magical flaxen-haired woman wearing twinkling lights and singing along to a synthesizer, perhaps. That happened to me once.

On the street, a Senegalese man wearing a Fila tracksuit wanted me to buy some fake transparent Balenciaga sneakers for eighty dollars.

I wondered how blockchain would affect this counterfeit market that never seemed to leave Canal Street. I didn't think it actually would. Since everyone who shopped the sidewalks of Canal Street knew that the luxury goods they were getting for such a good price were fake, why would any of them need to check provenance on the blockchain? Unless, I thought, the salespeople were able to make rare digital images of the counterfeit goods. That would legitimize them, as subjects of rare digital art, and give them true value.

In early 2018, Steve Bannon, former chief strategist for President Donald Trump, announced that he had invested heavily in bitcoin and other cryptocurrencies. Liberal utopians everywhere shuddered.

In other right-wing political news, after the violent white nationalist march in Charlottesville, some mainstream institutions, including PayPal and some banks and social media companies, quit working with avowed racists. Many of those racists then turned to cryptocurrency as a means of exchange—neo-Nazi leader Richard Spencer had already declared bitcoin the currency of the alt-right. Soon, a Twitter account called @neonaziwallets began posting the bitcoin activity of thirteen different bitcoin accounts it had deduced were linked to pro-Nazi, white nationalist groups, exposing their financial activities to the world. This curious mix of a free-flowing, no-intermediary economy and perhaps unwanted transparency is bitcoin in a nutshell.

Curiously, it seems that blockchain and bitcoin might also have the effect of encouraging more equality, by offering access to financial markets to people of color, and those with lower incomes who have traditionally faced more barriers to entry to the investing class. Blockchain technology can help level the playing field so that people with limited resources from all social groups are able to invest and participate in the marketplace in ways they've long been denied. Fractionalization of ownership can make even the smallest investment valid, especially if government regulations keep pace with the evolution of blockchain.

Another equalizing idea that is getting a great blockchain boost is the movement for universal basic income. An innovative and sometimes puzzling Berlin startup called Circles lets people create their own individual cryptocurrency to use in what they call a postcapitalist economic system based on collaborative interaction.

When someone joins Circles, a smart contract mints him or her a certain amount of crypto out of thin air. This is hard for most people, including, at first, myself, to grasp. An algorithmic proof of work protocol regularly adds more of their personal coins to their digital wallet. The value for this crypto arises from the fact that all the users of Circles see each other's currencies as equally valid. They can accept currency for a service they provide, and they can spend it in stores that participate. Circles now operates at a local scale, with shops and restaurants in Berlin accepting the currencies. It's expected that as more people join, all of the individual Circles currencies will meld into one giant shared currency, perhaps even a global currency. In this economy, everyone would have a basic income as new crypto would be regularly created for each person. Of course, there are many potential pitfalls, including the real specter of inflation. As with so much involving blockchain, the proof will be in the intangibles.

In Atlanta, and elsewhere, a growing black-tech movement is using blockchain to address issues that affect the African-American community. Atlantian Ed Dunn has a fascinating project to use blockchain to establish a free market exchange program that's intended to provide all kinds of services for black Americans, although it seems that anyone would be able to access the products, from repurposed clothing to investments.

Blockchain is uniquely designed to level playing fields, because it does not have to be controlled by the institutions that often create the problems that lead to inequality. As an unemotional ledger technology, the blockchain underlying all dapps is blind to ethnicity, color, and class. All over the world, it's giving power to those who need it the most.

Forests are vast, complex distributed systems, where individual trunks are joined into a community by roots, pheromones and other chemicals, sounds, and perhaps other communication tools as yet undiscovered by humans. Trees use these systems to warn each other of dangerous chemicals in the environment, and to share nourishment when needed. However, given their stationary, nonviolent nature, trees can't harness their distributed systems to protect themselves from loggers intent on clear-cutting forests.

The terra0 framework, on the Ethereum network, is meant to help them out. It's hard to say whether it is an art project or a forestry innovation, but terra0 is designed to turn forests into DAOs, meaning they manage themselves by selling timber and acquiring more land. In effect, the forest owns itself.

Here's how that happens:

You decide you want to help a forest take care of itself and expand in a sustainable way. You buy some forested land and establish a distributed network of trees equipped with sensors that monitor tree growth, density, human and animal visitors, and other measurements. You create a system of smart contracts to regulate all of this for the forest DAO. By this point, you have given up ownership of the forest, beyond your participation in the DAO. The forest enterprise does an ICO to raise money, based on the future value of its wood and other resources. You receive a share of the coins, along with all of the other investors, and it becomes autonomous. The tokens give the forest capital to earn more money. Smart contracts and other dapps built on top of the chain allow the forest to sustainably harvest itself (by hiring human or robot foresters), cultivate and protect other plants and species, and invite humans to visit, for a fee. Eventually, the value of the tokens rises to a level that makes you want to sell them and use the money for another purpose. The forest might buy them from you, if its algorithms think that is right, or an outside investor might want to be part of this DAO. The forest continues earning money, saving what it can to purchase more land and expand its boundaries. This creates a self-sustaining, well-managed forest that holds value just for being a forest. There's no need for it to become condominiums or a ski resort, or to be exploited in any other way.

I think many people periodically grow tired of technological promise, of the suggestion that the future might, somehow, be brighter this time. Understandable. But I urge you not to give up quite yet. Look at how the oldest blockchain dapp, bitcoin, is helping people see more clearly in sub-Saharan Africa.

I spoke to the director of a U.S. charity that provides glasses to people in need around the world, and that often has problems receiving payments and transferring money, due to cumbersome rules. For instance, its Nigerian partner has a difficult time converting that country's naira into dollars and sending them out of the country. Recently, the partner has started buying bitcoin with naira and transferring those directly to the charity in the United States. On receipt, the charity in the United States exchanges the digital coins for dollars. The fees are minimal. There's no attempt to profit off the coins. It's just a way to use blockchain to avoid the government and financial mediators that make transferring naira to the United States very difficult.

STACKS OF APPS CALLED DAPPS • 135

Who owns an image? Looking at a photo of a Texan with an AK47 slung over his shoulder, I wonder what he'd do if he knew I had this photo. He's standing in the crowd at an antigun rally, intimidating the protesters and giving the cops extra work. Since it's an open digital file, I can do anything I'd like to with this image (which in no way is my property, although I can act as if it is). I think I'll post it on my blog.

It's so easy to copy and paste or embed someone else's work into your own website that many people have come to see this appropriation as a right. To prevent such activity, you must be able to prove that the image is yours, and better yet, demonstrate that you copyrighted it. Visible watermarks are one solution to declaring ownership, but they mar the image. Metadata embedded in the image goes a long way toward protecting the originator's rights, although there are workarounds. Fortunately, a number of new services let you use distributed technology to easily take care of copyright issues.

I spend about two minutes registering my name on the website of a company called Binded and uploading the photo of the gunman. The people at Binded direct me to lock the image into the blockchain, using a hash to create an immutable record of the time and day that I claimed ownership of this photo. They'll file the copyright with the U.S. government. Forever after, their algorithms will scour the web to find anyone who might be using the image illegally and ding them for damages. Not bad for a few seconds of work.

It's a nice shot. And now I "own it." The only problem is that the photo wasn't mine. My brother, Peter, who lives in Austin, emailed it to me earlier today. He took the photo of a counterprotester at an antigun rally with his own camera, while I was 1,500 miles away. My "ownership" of this photo is a reminder of the blockchain's limitations, and the fact that an intellectual property registration is only as truthful as the info that is put on it.

However, if my brother had registered this photo when he took it, there wouldn't be an issue. A quick search of various blockchains would have revealed that he was the first, and that my ownership claim was fraudulent. Good thing for me that he didn't do that.

The sex industry often leads technological innovation, from the humanoid sexbots that staffed brothels at the World Cup, to tiny voyeuristic camera drones that can hover outside bedroom windows unnoticed. Yet, so far, blockchain-based sex services seem pretty tame, and sometimes even earnest. The cough-syrup-colored illustration for the LegalFling phone app shows a sexy woman grinding on a hairy-chested man as he taps on his phone screen to send her a consent request. As the company describes it:

> *During a fun night you meet your fling. Now it's time to get consent. Does your fling really want to take it further? Simply open the LegalFling app, scroll to your contacts and send a request. Your sexual preferences, including your do's and don'ts are automatically communicated.*

Maybe he likes missionary and light spanking followed by a bagel with a schmear. The app reveals these boundaries to her, so she knows what she is getting into.

Will she tap yes on her phone, immutably committing her consent to have sex with him to the blockchain, so that he can cite it should she later claim the encounter was nonconsensual? Or will she knock the phone out of his hand and leave the apartment immediately? Awkward, right? And what if she taps yes, but then decides "Nah, I'm not interested"? No still means no, even if you've already consented, says LegalFling. Which kind of renders the whole thing meaningless.

Another dapp, called Intimate, offers porn vendors, sex workers, and those who pay them a private way of exchanging money, far from the prying eyes of police, church elders, and spouses. The idea is to have Intimate serve a variety of vendors, so that a person's privacy and reputation are recognized across all the stores, sex workers, and websites they visit. Payments are done via smart contracts, using cryptocurrency and anonymous addresses. Intimate says this confidentiality makes it safer for all parties than using standard websites and credit cards.

An incipient company called Exo seems to have hit upon its own blockchain formula that could lead to good sex, privacy, and fun, especially for lovers who are far apart. They've been developing an Exo Suite of touch responsive and sensation responsive sex toys (called haptic and somatic toys), including the ExoWand, ExoThrust, ExoTouch, and ExoSuit. A person alone in a hotel room, for instance, can link through the dapp to another person far away who has his or her own Exo Suite, and they can stimulate each other and feel their partner's response. Push a button, and your lover's ExoThrust vibrates. Wave your ExoWand and your lover feels the movement. While all of this could probably be accomplished right now, using the regular old Internet, blockchain adds the crucial element of privacy that's missing from most online porn, sex toy purchases, and sex worker exchanges.

Hunting Unicorns

If everything seems under control,
you're not going fast enough.

—*Mario Andretti, race car driver*

The Byzantine Empire was the continuation of the Roman Empire into the Middle Ages. The capital, Constantinople (present-day Istanbul), was dominated by Greek, rather than Roman, influences. The empire fell to the Ottoman Empire about forty years before Columbus discovered the New World, and it never recovered, making it the perfect imaginary location for a parable that would become central to the evolution of blockchain systems.

The Byzantine Generals' Problem is the most often cited explanation for how a distributed computer system can be resilient enough to remain reliable even when some nodes don't function well. A suitably resilient system is called Byzantine fault tolerant. The parable goes like this:

A Byzantine army is led by a group of nine generals, each of whom has taken up a position at a different point around a walled city they'd like to control. The city is well fortified to defend itself, and the generals have the choice of attacking, at great peril, or turning tail and suffering ignominious defeat. To take the city, the generals would have to attack in unison, according to a common battle plan, from their separate positions. If any one of them were to take a different approach, surely the entire army would perish. However, the generals can't gather together to discuss the situation, but must instead rely on messages to each other by courier.

Unfortunately, one yet-to-be-identified general is duplicitous. He might not vote for the best plan, or he might even purposely undermine the others. If four of the generals want to invade at sunrise, and four want to invade at sunset, the remaining general could choose to send a courier to the first four supporting their decision, but also send a message to the second group supporting their decision. This would assure chaos and failure.

This scenario applies to distributed networks, where various types of computers or nodes (the generals) are linked (the messengers). In order for the network to function honestly, there must be a way for all of these nodes to agree about transactions, data, and other information that enters the system. With the generals, if those loyal generals who wanted to attack at night held the majority, then a state of what computer scientists call Byzantine fault tolerance would be achieved. The other loyal generals would agree, and the system couldn't be undermined.

With computer networks, this is done with consensus protocols that allow the entire group to certify the authenticity of the data that is organized in blocks. Not to ensure that the info is "true," but rather to certify that it was entered properly, and was entered on such and such a date. There are a number of ways to do this. The most common is called "proof of work," and it's the method the bitcoin blockchain uses to find consensus and certify transactions.

Proof of work is done by the nodes of powerful computers that "mine" for bitcoin. These are often large collections of computers linked together in industrial buildings next to hydropower or other inexpensive sources of electricity, because the processing power needed to compete in bitcoin mining uses up huge amounts of electricity. The more computers you have, the more likely you are to solve the algorithmically generated math puzzle that allows you to certify that a block is prepared well and closed—the proof of work protocol. In the early days of bitcoin, there was so little competition that a person could mine coins using a desktop computer. Each time a node solves the mathematical problem needed to close a block, the algorithm rewards the miner.

In the first years of bitcoin, a miner would earn fifty coins plus a transaction fee for each block. That amount was programmed to drop by half after every 210,000 blocks were confirmed in the chain. When the coins were worth only a few dollars, there wasn't much mining competition. Now that they are worth thousands of dollars each, depending on the market, professional mining companies have established themselves. China leads the way, but there are also mines in other countries around the world. Some are set up in cold countries like Iceland, so money won't have to be spent on air-conditioning to cool the mining computers, which generate tremendous amounts of heat. Every time a block is ready to be certified, these miners compete to see who can solve the algorithmically generated math problem first, and collect the reward. It's tedious but very rewarding work. In early 2018, the reward of twelve bitcoins would have been worth nearly a quarter of a million dollars. Once a miner certifies a block, the proof is sent out to all the other nodes on the chain, who approve it. Then the miner can sell the coins on one of the many exchanges or markets to be found all over the web, or hang on to the coins as an investment.

Roam the landscape of innovation and you're bound to hear the warbly call of Silicon Valley venture capitalists hunting easily spooked unicorns, and the deep-throated roar of social innovators wanting to bag a cure for the world's ills.

Scale. Scale. Scale. It is an obsession of our age. Everyone wants to blow up in unprecedented ways. Many, though by no means all, experts believe blockchain technology might answer those calls. A variety of new protocols suggest that the blockchain might be able to scale to meet the demands of any enterprise or population, no matter how large or complex.

In theory, most open, distributed systems, like blockchain, are ideally structured to scale larger, infinitely. The more nodes on a system, the more resilient it will be to failures within the system, intrusions and attacks, and resource shortages. The network effect will ensure the system grows exponentially. Blockchain would seem to fit the bill. However, its architecture is not eminently scalable—yet.

Right now, growing by leaps and bounds is difficult, because every node on the chain in effect processes, and stores, every transaction. The system is secure, yet cumbersome. In early 2018, bitcoin could process about 3 transactions per second, using proof of work consensus. At the moment, Ethereum processes about 25 transactions per second, using a mixture of proof of work and proof of stake, which it is introducing slowly into the mix, so it can monitor security and other issues. That's a tortoise's crawl, compared to systems like Visa (25,000 per second) and Swift (more than 300 per second) and Mastercard's centralized system, which can handle 44,000 per second. But we all know not to count out the tortoise.

Speed is vital to scale, but a blockchain is valuable only if you can retain its immutable ledger qualities, which are maintained by the various consensus protocols. To scale to the point where it can handle multitudinous transactions, such as for a financial services firm, there has to be a more efficient system that doesn't require approval from every single node on the chain. Somehow, the work of consensus has to be divided among the nodes.

The Ethereum blockchain, founded by Vitalik Buterin—that willowy Canadian, who was nineteen at the time—began using proof of work, as with the bitcoin chain. As Ethereum has grown into a technology supporting smart contracts and other innovations, the demands of scale have led to congestion. For instance, for a few moments in 2018, those tradable CryptoKitties were so popular that the demand dramatically slowed the Ethereum chain.

In 2018, Ethereum began testing proof of stake, which uses "validators" in place of the "miners." Miners race to be first to perform the algorithmically generated math problems that let them validate the blocks of data, thus earning bitcoins for their efforts. This is a cumbersome and energy-intensive way of closing blocks, and it's hoped the validators will make things work better. To be a validator, you deposit a portion of the ether coins you own into an escrow. You are "betting" this money on the fact that you think your processing system will snare the next block coming down the line, so you can verify it. If you are the first to verify a new block, then you get your ether back, plus an equal amount of new ether as your incentive for participating. If you are unable to certify the new block, you lose your escrow.

This system is expected to greatly reduce power consumption, which is an estimated $450 million-a-year cost for the bitcoin blockchain. It also is designed to level the playing field, because you don't have to have specialized computer mining equipment to participate. It promotes good behavior, in keeping with the ethos of the system. Most important, it speeds up the chain, allowing blocks to be certified and closed much more quickly. The issue of releasing coins, through mining, validating, or other methods, is an area of great uncertainty for blockchain. It must be solved if blockchain is to succeed.

The bitcoin proof of work protocol has been heavily criticized for ensuring that the bitcoin network uses enormous amounts of electricity—on a given day, according to some, it consumes as much energy as Denmark. It's been estimated that the power used to process just one bitcoin transaction could run a small house for a month. That level of waste is unethical to me. For this reason, I have never purchased bitcoin.

However, there is another way to look at it that might be more relevant in the future. If bitcoin were to scale to a level where it began to make banks obsolete, it might actually be more sustainable than our current monetary system—which also consumes vast amounts of power. Think of how many bank buildings and ATMs there are worldwide, all burning electricity 24/7 to power lights, air-conditioning, and machines. If bitcoin and other cryptocurrencies replace this system, there would be little need for these buildings, and the energy they consume. I imagine that bitcoin might start to seem energy efficient in comparison to the energy used to run banks now.

Still, proof of work is not reasonable, or useful, for many nonbitcoin applications. To that end, many blockchain geeks are working on new protocols that should both speed up the transaction rate and reduce fears of environmental damage from proof of work operations. If blockchains are going to take over the world, they need to work efficiently and well.

The consensus protocol space, including proof of work, proof of stake, and other methods, is exploding with new ideas. Hashgraph, which is not exactly blockchain but lives in the same neighborhood, seems promising. Its consensus method is called Gossip, and each node randomly chooses another node for authentication. IOTA is another, using a system called Tangle, though it's faced a lot of criticism for being insecure. There are many other blockchain and pseudo-blockchain projects in the works.

Passionate visionaries and engineers around the globe are pushing distributed ledger tech as fast as they can. It's impossible to predict which tech will win out, but it's a fair guess that things are going to improve rapidly.

Many believe that an innovation called sharding will save slow and overloaded blockchains. Developers on the Ethereum chain have been working on this technique. In effect, sharding divides the blockchain into sections that occupy different nodes, in order to keep the traffic through nodes lighter. It's a huge shift in thinking about distributed networks, partly instigated by a group called Project Chicago, that gets intellectual and practical—at the same time—about all the "commodities" (transaction fees, storage space, and more) used in determining the value of cryptocurrencies. Developers there, and at other organizations around the globe, are working on sharding and other techniques that will improve the movement of value on blockchains. The most radical among them believe that this distributed effort will result in a revamping of the entire blockchain ecosystem.

The new distributed Internet. Internet of blockchains. The free Internet. Your Internet. The perfect Internet. The Internet that will finally not need to be written with a capital "I." The Internet we all wanted back in 1994. The Internet the Grateful Dead would have sung about, if it had happened in time. The Internet that houses the Internet of cool things.

This Internet does not yet exist. However, many people are trying to scale it into being.

What if all 1,600 or 2,000 or however many types of cryptocoins in existence at this minute could all communicate with one another, so that you could spend your type of coin in another coin's ecosystem? Say you had laughiecoins, and you wanted to buy stuff from the store running on depressocoins. You could do it, seamlessly and without penalty. That would be cool, like a foreign currency exchange that didn't require numerous middle agents and fees to transfer your money. That would be the Internet of blockchain at work. It's not happening yet.

Say you are tired of how the Internet now seems largely dominated by corporations, including Facebook, Google, Netflix, the *New York Times*, and others. You crave a new, liberated system that doesn't track you or take control of your data. Well, that's the new distributed web. Instead of web pages and other info being stored in the cloud, it will be stored on everyone's phones and computers and dishwashers and cars. You'll be paid to run a portion of this new Internet through the unused processing and storage capacity on your devices. (The fact that it will burn up all your battery time has not yet been addressed.) This Internet, too, is not here yet, but many cool and smart people think it's on the horizon. As soon as you notice that people have stopped saying "the new distributed Internet," and have started just saying "the distributed Internet," you'll know it has arrived.

I believe that will be soon.

Ummm, one little thing: power failure.

This technology, like most of modern life, is based on easy access to an endless stream of electricity that now flows seemingly effortlessly into our homes and businesses, cars and data clouds. We put tremendous trust in our electrical grid, even though most of us have no idea how it works, or how fragile it might be.

I was shocked when Hurricane Sandy knocked out power in my New York City neighborhood for five days. Along with many other downtown residents, every morning I'd walk to Midtown to find a shop or diner where I could charge my phone, which I used as a flashlight when night came. This was my third New York City blackout—it definitely can happen. What happens when a wide swath of the country loses power?

Even those who generate their own renewable solar or wind energy are usually attached to the grid for backup. The grid is threatened by hackers, nitwits shooting shotguns at substations, and weather. More than weather, even, squirrels threaten the grid. They like to chew through wires, build nests on transformers, and otherwise wreak havoc. Don't believe me? Check out CyberSquirrell.com.

If the grid goes down, our blockchains will stay alive on all the nodes that generate their own power or are located beyond the edges of the affected grid. The record of the blockchain can survive in a single idle hard drive. However, if the situation is worse—cascading outages that continue to spread across the continent—then our chains could get pretty weak. This is when the old-fashioned sense of distributed living will start to seem attractive. At that point, good old distributed mysticism and faith in God just might come in handy as blockchains collapse.

Self-Sovereign

Sovereignty is not given, it is taken.

—*Mustafa Kemal Atatürk*

Recently I went to an orientation for new students of an art school. At the sign-in table, next to the name tags, were stick-on ribbons, each one printed with a word or two: jock, he and him pronoun, withdrawn, they or them pronoun, gay, she and her pronoun, LGBTQ, studious, cis male, cis female, of color. Some students tagged themselves with three or four of the ribbons, like generals back from a campaign.

We live in the age of identity, of course, and are constantly asserting our identity, whether it is a political or philosophical affiliation, a gender, a race, or a nationality. We demand respect for who we are and what we believe. However, for the most part, we are not in charge of our identity. Banks are. Governments are. Schools, churches, clubs, and fathers are.

Freedom, equality, a level playing field. We hear these three desires from all corners. Freedom is where the right wing and the left wing meet, around the back of their edifice of opposing ideas. Likewise with equality. The disenfranchised "forgotten" white middle American seeks parity with the elites on the coasts. Black voices demand equal recognition with whites, forgotten or not.

For me, the heart of blockchain's incredible potential is its ability to make all humans as independent as the day they were born, and the day they will die. And in that independence, have the ability to connect to their own constellation of other independent people. While blockchain technology allows business and governmental transparency, cooperation among strangers, efficiency, environmental progress, and much more, the foundation of all of these advancements is the technology's unique ability to let individuals own their identity, trust the identity claimed by others, and make decisions as they see fit.

The term "self-sovereignty" comes up often these days in the blockchain world. Just as a king is sovereign of his country, with blockchain, any individual can be sovereign over his or her own identity. This would include sovereignty over the data a person produces. Right now, we give that data away to Facebook, Whole Foods, the utility company, the government, and many other entities. With a blockchain-based self-sovereign identity, any person would be able to control the release of data about his or her purchases, travel, and musical tastes. For instance, if Amazon wanted to know who was buying solar-powered garden lights, it could get the data, for a price that would go to each individual.

To accomplish this, you register your name, date of birth, citizenship, education, marital status, race, driver's license, and anything else you'd like to use as an identifier, on a blockchain. Software built by several companies, including uPort, makes this process fairly easy, at least in theory (I found uPort buggy and challenging; I'm sure that will change over time). You support the veracity of the info with documents, or attestations from the hierarchical institutions that certify us, such as the county that certifies your birth record. But once everything is certified and on the blockchain, the info can't be questioned.

The key to all of these efforts, and to any widespread program to put everyone's ID—including the IDs of people who currently have passports and driver's licenses—on a blockchain, is to let smart contracts and other methods give citizens control over what pieces of information are released and to whom. Privacy is key. For instance, you might not want the clerk checking your ID at 7-Eleven to also have access to your financial records.

Here's an example of how self-sovereignty on the blockchain might change the world: An ID registry on a blockchain could transform the lives of refugees across the globe.

A few years ago I visited a Congolese refugee camp on the Rwanda side of Lake Kivu, just across from the Democratic Republic of the Congo. The camp had been in place for more than thirty years, and I met a man who had been born in the camp and still lived there, now with his own wife and child. The long-term residents lived in mud-brick homes along narrow lanes. These packed-dirt walkways led, eventually, to a clinic, a market run by local villagers, a DVD-fed cinema run by an enterprising refugee, and public bathrooms for all 30,000 residents of the camp. Because the camp had no electricity other than a couple of generators that wealthy residents used to earn money, the lanes were dark at night, even with a moon. Often, women were assaulted as they made their way to perform their most basic tasks. The refugees had escaped violence in their home country, and now lived in violence and poverty still, under the auspices of the UN.

The economy of this camp was completely based on barter and black-market activities, with the residents often selling a portion of their UN food rations to outside traders who set up a market within the camp. Many of the refugees had no passports or IDs other than what the United Nations High Commission on Refugees had provided for them. In effect, they were stateless, nonexistent from a bureaucratic point of view. The World Bank estimates that more than 1.1 billion people on earth, like these refugees, lack identification. That number includes at least 3 million U.S. adults who lack ID because they don't drive and can't afford a passport. Being without ID in this world means you can't open a bank account, get credit, cross borders, or get married. In some U.S. states, you can't even vote.

But things are changing. The majority of people without IDs now have cell phones or can borrow one. With these devices, they have access to blockchain technology, which can give them permanent, immutable IDs. These records would allow refugees to enter financial markets, apply for citizenship, and exist more fully in the world of nation-states. The IDs give them reliable access to digital services on the Internet, without having to turn to gatekeepers such as Facebook.

The UN is working with Accenture and Microsoft to make a blockchain ID system called ID2020, which will help achieve one of the UN's sustainable development goals of providing ID to everyone in the world who needs one. In Moldova, rural children are getting blockchain-based IDs as a way to prevent human trafficking, which is a problem there. Retina scans connected to a blockchain database have been used to authenticate Syrian refugees using charitable food donation cards provided by the World Food Program. In New York, an enterprise called Blockchain for Change is trying to provide workable IDs to homeless people and others.

And those refugees I spent time with in Rwanda? If the UN succeeds with its projects, perhaps in the future the UNHCR IDs the refugees receive when they enter the camp will be permanently registered on a blockchain and serve as a valid ID when—if—they are allowed to leave.

The identity of voters is increasingly being questioned in the United States, and has long been an issue in other countries. In May 2018, a few dozen West Virginia citizens who were deployed in the military, some overseas, used their thumbprints to verify their identity and proceeded to cast votes in the state's primary elections. The votes were cast using smartphones that registered them on a blockchain. This limited test pilot, using a system developed by Voatz, was to determine if more voters could use their phones and blockchain technology to vote in the future.

———————————————

There are several potential advantages to voting online, in a blockchain system.

- A person's identity is verified through biometrics, such as thumbprints, or other methods. This erases the possibility that people might vote more than once.

- Votes are permanently recorded, and people can always check to see if their vote was recorded correctly. No more lost boxes of ballots, or a fear that someone in a back room is messing with the votes.

- Miscounts are impossible. For instance, the "hanging chad" dispute that plagued the 2000 presidential election between George W. Bush and Al Gore would not have happened had all the votes been clearly recorded on the blockchain.

- Voting is easy. No need to go to a poll and wait in line. A person could vote from the far side of the world.

- Identification questions are nonexistent, as ID will be based on a fingerprint or other immutable data.

- Absentee voting can be done anonymously. Currently, anyone submitting an absentee ballot via email or fax gives up their right to privacy.

- Security is guaranteed, because each vote is visible on the blockchain, although the person who voted remains anonymous.

- Auditing the vote is simple, since the chain is transparent.

- Results are available instantly.

While democracy on the blockchain is an enticing idea, implementing the technology in a vast country like the United States would be difficult. It's more likely to take off at the local political level, and in boardrooms, university committees, and other places that can more easily adapt to change, before spreading to national politics. If it does, then election errors and voter suppression could become a thing of the past.

In one of the first tests of blockchain voting, in 2016, the country of Colombia gave expatriate Colombians the opportunity to cast a symbolic (uncounted) vote on whether to approve an important peace treaty between the government and a rebel group, Fuerzas Armadas Revolucionarias de Colombia, or FARC. In addition to allowing a "yes" or "no" vote, the *Plebiscito Digital* (*Digital Plebiscite*) ballots that appeared online, on phones or computers, offered "subthemes" about specifics regarding land redistribution, narcotics trafficking, and other related topics that the expatriates could vote on. These questions were called "liquid Democracy," because they were less rigid than straight up or down votes.

While overall the vote was a success, the organizations that sponsored it discovered several problems with blockchain voting, at least among expatriate Colombians:

- The tech wasn't well enough developed and the interfaces weren't easy enough to use.
- Some people didn't have high-speed Internet, or great computer skills, and might not have voted at all.
- There was a lack of support from political leaders and organizations.

Still, it was a remarkable early test of what the future of voting could hold.

The New, New World

Everything that can be automated will be automated.

—*Robert Cannon, senior counsel, the Federal Communications Commission, 2014*

Blockchain will destroy brands. Colgate, Uniqlo, Apple, and all the others owe their strength to an ethereal authority invested in their brand. What does Ralph Lauren really mean to its customers? American luxury. The freedom of sport and the West. Better-than-average workmanship, for sure. What would the brand mean if everyone could see where its cotton was farmed, and by whom? Who sewed the clothing? What fuels were used to transport the finished goods across oceans? How much profit the company earns on that indigo vest? Depending on the answers, this information could help the brand, or damage it.

Most brands work from an entrenched sense of privacy and a need to control their inputs and outputs. With brands we like, we trust that they both know what they are doing and want to share themselves with us. In the coming years, we may have very concrete facts on which to base our decisions.

As it evolves, blockchain tech could expose many of these tightly managed brands to previously unimaginable scrutiny. For some brands with dark secrets, the transparency, immutability, and democratizing effects of open blockchains could shatter the trust they've been building for years.

Why would brands subject themselves to such scrutiny? We are facing a sea change in manufacturing, selling, and consuming. My bet is that transparency is going to become the norm, and that this will affect our worldview in ways we can't foresee. Companies will become transparent, or disappear into an opaque obscurity.

While there is much debate now about whether open, or permissionless, blockchain technology will prevail over private chains,

I believe the public chains will be more useful in the long run for businesses wanting to innovate. The more open a company is about its materials, sources, movement of goods, payment practices, and treatment of consumers, the more creative it will become. With transparency, great thinkers who might otherwise be left out of the idea stream will come to the forefront. Companies that are open to "distributed" creativity will do well.

Consumers will see this. They will develop new metrics for trusting established brands. Brands that aren't open will be perceived as inherently less trustworthy. In the future, consumers will in many ways be less "manageable." The consumers will become the managers.

What's more, given the potential of smart contracts on the blockchain, it is likely that soon, new companies will appear that are owned by thousands of participants and run largely by robots. Rather than responding to a centralized authority or board of directors, everyone who participates in that blockchain will have agency. The inmates will have taken over the asylum, and that might be the best thing for all of us—especially those who let their brand evolve with the chain.

Honesty will be the new mystique.

The communist blockchain thrives in Shanghai.

I arrived in Shanghai aware that China had upended the world of consuming and spending, with services like thirty-minute guaranteed delivery of everything from live Alaskan king crab (picture that on the back of a motorbike speeding down a sidewalk in the French Concession) to a bag of spring onions. Yet, because my daughter had been living in the city this year, I was also well aware of the technological limitations she'd faced using China's ubiquitous WeChat and Alipay, which everyone from the granddad smoking on the sidewalk to the fintech (financial technology) analyst living in a Pudong high-rise uses to pay for most goods and services.

Most stores and kiosks hang QR codes protected by Chinese Saran Wrap in front of the cash register, so buyers can scan the payee information with their phones as they wait in line. Usually, payment is a breeze, but every fourth or fifth customer, it seems, has to pass his or her phone to the clerk for some frantic app juggling while people in line shift impatiently. As I stood in a thick crowd to board a high-speed train, a woman thrust her phone into the air, jabbing it higher and farther, moving here and there in a desperate search for enough signal to download her ticket.

Problems are compounded for foreigners. My daughter didn't have a Chinese bank account, and could take advantage of only some of WeChat's payment offers, even though she used a Chinese phone. I paid in cash most of the time, which felt archaic even outside Shanghai. When I asked to change an airline ticket, I was told the refund could only be made to a Chinese credit card—I just let that fifty dollars go. There is mistrust built into the system, especially when it comes to foreigners.

That makes urban China a perfect storm for blockchain break-throughs. Chinese innovators are among the most heavily invested in blockchain technology (including bitcoin mining, which seems to be under almost monopolistic control of Chinese enterprises, for better or, say many, worse). The automobile company Wanxiang recently announced a $30 billion investment over the next seven years in blockchain for smart cities. Wanxiang is building a massive lab in Hangzhou, an hour by high-speed train from Shanghai, to promote blockchain innovations. That's on top of a large venture fund focused on blockchain tech. It's a natural step for an auto company that's facing a future of self-driving cars and DAOs.

Clearly, China is poised for liberation by blockchain. Whether the state will be able to shape the unknown forces blockchain will unleash is far from certain. Already, the government has banned ICOs. The government's powerful social credit system, which is expanding to track the behavior of each citizen—everything from criminal behavior to whether a person takes care of elderly parents—and reward "good behavior" with access to better trains, schools, and other services, could be undermined by blockchain's protection of personal data. Or enhanced by its ledger-keeping qualities. No one is sure. Given the country's rapid evolution into a mobile payment and banking econ-omy, and the scale at which business takes place there, all eyes should be on China, particularly Shanghai. Both the utopian and dystopian potential for blockchain will be on display.

Terrified by AI? Don't be. Artificial intelligence is far superior to human intelligence in one respect: AI doesn't worry—unless you tell it to. However, most humans worry a lot about AI. Even Elon Musk is freaking out about it. Artificial intelligence and blockchain together are going to have huge, as yet unpredictable effects on our lives. Let's turn our backs on the naysayers, and look at the good.

When medical records migrate to blockchain, AI can be used to analyze anonymous versions of these oceans of data to understand disease and figure out treatments. AI could also be used to continuously examine copyright and trademark data on the blockchain to spot violations.

Imagine an electric utility using AI to negotiate the smart contracts it builds into blockchain transactions on its grid. That AI would be tasked with maximizing profit for the company, through efficiency and price, using smart contracts. Meanwhile, a small solar energy producer—even just a homeowner with solar panels on the roof—would have its own AI to negotiate with the utility's AI. All of the transactions would be recorded on the chain. Machine on machine, each learning as it goes, all the trades facilitated by trading bots using smart contracts on blockchain.

Sounds crazy, for sure. Well, the Wright brothers sounded crazy, too.

Approaching Dubai from above the Arabian Sea, you descend over Oman and a vast arid plain. It seems impossible that the 160-plus-story Burj Khalifa could be ahead, because below there is nothing but red rock and gray sand. Soon squares of cultivated land and blocks of gray rooftops appear, then, in the distance, the skyscrapers of this city rise improbably along the shoreline, like an apparition of modernity. You land, and it's all true. This city feels like the future rising from pre-history. On the ground are electric cars and men in dishdasha and ghutra, women in abaya, Shake Shacks, and camel-hitching posts. It's an improbable manufactured place, with man-made islands in the shape of palm trees, and an elite that's always pushing for more. Most of the infrastructure has been built just since the turn of the century.

This maddening and innovative capital of the Emirate of Dubai, a state within the Arab Emirates, is poised to have the world's most extensive governmental blockchain platform. The stated goal is to be the world's "first blockchain city" (the definition is open to interpretation) by 2020.

Blockchain as a service is a linchpin of Dubai's goal to be a "smart city" that's competitive globally and good for its citizens. The government estimates it will add tens of billions of dollars to the country's GDP in the next few years. The effort will include autonomously run transportation systems, a focus on sustainable practices, extensive IoT connections, AI, and government services that are 100 percent digital and accessible to all. Of course, blockchain plays a key role. The vision is for cashless, paperless interaction between the government and the people. Dr. Aisha bin Bishr, the director general of the Smart Dubai Office, has said that one goal is "making Dubai the happiest city on earth, spreading our happiness through our global community."

Blockchain is the fever dream of radical thinkers, as the potential of smart contracts, distributed nodes, and a new financial system encourages mad speculation about what might change. Given blockchain's potential to enhance markets, profits, and financial wizardry, while undermining the status quo, the big traditional players, such as Goldman Sachs, are going at it hard. There promises to be an epic shakeup of the technology as hierarchical, more traditional systems, such as Wall Street firms and hedge funds, embrace the chains, and distributed, almost utopian companies, such as ConsenSys, battle those firms for hearts and minds. We do not know how this will turn out. There could be several new systems, and maybe they'll work together.

My fantasy is that radical change might occur as distributed decision-making on the blockchain makes powerful institutions obsolete. If the idea seems improbable, remember that there used to be a Blockbuster video rental store in every neighborhood; that newspapers were printed in several editions a day; that film cameras ruled the universe; and everyone carried a lead pencil and eraser. Our most trusted institutions fall quickly in the face of innovation, especially unanticipated innovation.

These institutions are particularly susceptible:

- Banks, as they lose their authority (and ability to charge us for every transaction).
- Electric utilities that crumble in the face of distributed energy and battery storage.

- Wall Street as we know it. Even if it embraces private blockchains, it will never be able to control the evolution of cryptocurrencies or keep up with the financial innovations appearing all over the world.

- Dominant political parties with statewide control. Get ready for microparties and much more cooperation.

- Media conglomerates that depend on control and exploitation of intellectual property for their profits.

- Health care conglomerates that monetize free patient data and manipulate care.

- Social media: Good-bye, Facebook, Twitter, and Instagram. You exploit our data. We don't want you. Hello to social media we control.

- Centralized corporations, even if they have decentralized worldwide operations. The all-powerful CEO will have to listen to everyone on the chain, even if they say, "We no longer need the guidance of a CEO."

Lots of things on Wall Street move fast. Money isn't one of them. Different entities use different databases, and it takes emails, texts, and phone calls to clearinghouses and banks to execute a trade. Each entity has to establish trust with the next entity. Three days for a trade is not uncommon, and the deal hangs in the balance during the wait, until everything is set. These delays are called settlement lag, which sounds like something you'd take care of at a liposuction clinic. Blockchain to the rescue, if the trustless technology can reduce the wait to just a few moments, as many believe. Some powerful banks, including J.P. Morgan Chase, Citibank, and others, have formed consortia to address the potential of blockchain. Financial technology, aka fintech, startups around the world are working furiously to capture the prize that goes to those who figure out how to remove the obstacles.

The mainstream players are desperately trying to keep their toes in the pool. What happens if value can be stored on blockchains, without the need for banks? Or if credit scores become verifiable through IDs that are stored on chains, without the need for credit agencies? If accounting information is transparent and public? One big effort being made by a few dozen banks is a group called R3CEV, which hopes to standardize the technology for fintech blockchains, so communication between banks will be easy.

So, what will you be buying today? A latte? An electric lawnmower? A faux crocodile purse?

Any idea where this product comes from?

Chances are that none of the above items are sourced from just one place. The days of local manufacturing, using local products, are long gone. The latte, at the very least, will be made from beans harvested on a farm (or several farms, perhaps on different continents) in the high tropics. The beans will probably have been roasted in a central facility, in Italy, or Portland, or at the coffee shop itself. The milk will almost surely be a blend from several farms within a few hundred miles of you. The lawnmower and the purse no doubt have vastly more complex supply chains.

As I've said, most of those supply chains are secretive, stitched together with paper and digital documents that aren't shared along the chain of suppliers. Generally, there are many layers between the manufacturer and the supplier who begins the chain, with little, if any, direct contact. Transparency is limited. It's hard for an untested supplier to earn the trust to join the chain. Directives and forecasts are sent from the top of the chain downward. This results in incredible inefficiencies of time, materials, and marketing, and many missed opportunities. The creative opportunities that come from unscripted interactions with other businesses and consumers are thwarted.

Many of us believe that blockchain technology can allow new chains to develop that are supplier-centric, and also good for the enterprise at the top. Adding smart, robotic contracts to the mix can automate payment and clearance transactions that now take up a lot of paperwork and, more important, time.

The shipping giant Maersk and blockchain innovator IBM are building a system using Hyperledger that might revolutionize supply chains everywhere. According to IBM, moving a container of goods, such as flowers, from East Africa to Europe right now can result in a couple hundred paper documents for sales, customs, shipping, and more. They've devised a system whereby everyone on the supply chain, including customs officials and shippers, enters their transactions on the chain. Every participant is able to see the movement (and costs) of the goods. This deters theft, increases efficiencies—including payment, so suppliers don't have to rely on "factors," or moneylenders, to keep up their cash flow—and encourages trust and innovation all along the chain.

This allows suppliers to anticipate demand, and allows everyone on the chain to see the provenance and cost of goods sold. A UK company called Provenance is developing blockchain tech to ensure that food, such as tuna, is what the label claims, tracking it from "hook to fork." Often caught by certified sustainable fishermen in Fiji, the fish are tagged on the boat and tracked through the system. This helps retailers and consumers see that no slaves were used to catch or process the fish, as often happens, and that it was harvested with an eye on maintaining natural ocean stocks.

Again, with blockchain, the more people involved, the merrier. Companies that are willing to engage and entice their suppliers to participate will flourish. While efficiencies are obviously good news for companies, I think there will be other, more profound results from blockchain entering the supply chain. The egalitarian nature of the tech is sure to change how consumers look at products, companies, and the urge to buy. This tech will encourage companies to compete to be more transparent and accountable. This will then enable people to look more carefully at the human and environmental costs of the goods they consume.

In the end, that will encourage consumers to shop with ethical considerations, as well as aesthetics and "need" in mind. This will be good for them and the environment, and good for business.

Back to that cup of coffee. If you have at your fingertips an easy way of knowing whether your latte is made from sustainable crops grown by farmers who earn a decent wage, won't you be happier taking a sip?

Plagiarism just isn't what it used to be, back when a fellow had to take out a pen and copy down sentences verbatim in order to commit copy fraud. I did just that, in the seventh grade, copying a cover profile of Charlie Chaplin in that week's *Life* magazine for my semester project. Much to my surprise, I was busted, finger-wagged, and forced to sit in the hallway for the duration of that class. I never did it again.

These days, a person can just copy and paste. Many writers build their stories, chapters, books, movies, and love letters to their spouses by making an "outline" of copied and pasted ideas from a variety of sources.

I started doing this myself a few years ago. To build a researched story, I'll often string together a series of facts from a variety of different publications, tweets, or websites. I scrupulously note when I've pasted something I did not write, usually by putting it into italics, or boldfacing it, always with a source. I never use another person's words, though I'll certainly feel comfortable being influenced by their ideas. As far as I know, I've never copied anyone by mistake. I do not know of an example where someone has stolen my words. However, just in case, when I'm done I'm going to record this book on a blockchain, so that no one will be able to dispute the day and time that I registered these words. If no one has registered these words before me, ipso facto, I own them.

The structure of blockchain makes it the perfect narrative technology. You have the protected blocks, immutable, like cars on a guarded train roaring through the badlands, secure from bandits. Each train car contains its secrets, one coupled to the next. The chain is at once transparent and hidden in mysterious code.

Each of us brings our own meaning to this technology. That meaning can be good, as in: Blockchain will change the world for the better—or bad, as in: Blockchain is a destructive scam, built by fools to fool other fools. I choose to believe that the story and meaning are good. I believe that blockchain, with its honesty, trustless technology, and egalitarian nature, allows great sensitivity and beauty to flourish, by affirming the storytellers' ownership of ideas and creations, and letting them expand, unafraid, in the open.

This open connectivity is a mechanical manifestation of spirituality; it encourages faith. Look at the systems of the ancients, the Sufis, the Kabbalists, the Yogis, and others, with their distributed cosmologies and points of light. This technology is the beginning of the next age of enlightenment, although of course we aren't there yet.

Recently I watched U.S. president Trump give a State of the Union speech, and I was struck by how the speechwriter stacked up story after story of heroism, tragedy, patriotism, and pathos in order to meet the president's goal of winning hearts and votes. The stories were linked in a chain that connected the president to his citizen audience. The connection was brought home by the deep visuals in the balcony seats: a young couple holding the baby they'd adopted from a drug addict; the North Korean refugee amputee who just happened to bring the crutches he'd used before escaping the hermit kingdom; parents weeping at the mention of their murdered children. It was emotional overkill, for sure, and to be honest, I felt manipulated and a little guilty for watching. But even I was almost pulled in by the story. That whole speech was a blockchain.

Blockchain as a medium for stories is, to my mind, cemented by the blockchain creation story, which is a marvelous fantasy of brilliant deduction and escape. Any technology that begins with such a tale deserves more stories in its future.

Any company or person involved with blockchain should think always about the story he or she is conveying with the use of the technology. If a company is going to talk about blockchain, then it needs to be involved with blockchain in a wholehearted way. That doesn't mean it has to embrace the technology if it's not ready. But it has to be committed to whatever level of exploration it's ready to engage in. Any attempt these days to just "latch on" to the "blockchain thing" will be doomed, in the end, by the destructive powers of disconnection.

Ty Montague is a fascinating thinker and doer, and a pal of mine. A few years ago he wrote a book titled *True Story* that advocates what he calls "storydoing." At co:collective, the innovation company he cofounded with another sharp doer named Rosemary Ryan, Ty guides businesses toward living their story, rather than just telling it. Don't just talk about what you do, do it, and let your actions speak as loudly as your words. Let your business and personal life embody your story.

Regarding blockchain, that might mean using the technology itself as a storytelling platform. For instance, registering a product on a chain, including its provenance and the materials and people who made it. Then encouraging customers to add their story to the data. If it's a jacket, they could say they wore it while climbing Mount Shasta. If it's an expensive fountain pen, they could describe the book they wrote with it. Every person who encountered that object in the future would be part of the story.

And that's a lesson for everyone who is trying to build a story on top of the blockchain. The hype now is huge, and just mentioning blockchain along with your mission is a surefire way — for the moment — to get a lot of attention. But don't be fooled by the flattery, because in the long run — and blockchain tech is going to have a very long run — your story is going to have to be true, visceral, and transparent.

Be honest as you tell the story of your company and blockchain. Be frank about your level of involvement in this technology, whether you're just dipping your toes into the water or you're taking a swim across the ocean. If you don't understand how it works, say so. If you're getting excited, that's great. Let everyone know. And if you, like most leaders these days, are just not quite sure where blockchain is going to lead you, embrace that uncertainty. Tell the world you are studying. Take a stab at explaining how the distributed network will affect your business's creative inputs, supply chain, and strategy. Be true. Be daring. Be centered. And most of all, live the story you want to create.

Imagine the potential of blockchain on the electrical grid. It's time to dream a bit, because electricity needs big ideas right now. Not since the early twentieth century have so many changes come to the electricity industry. As we move into the new energy future of renewables and the Internet of Things, the grid faces significant changes that will forever alter the relationship between energy suppliers and consumers. Blockchain technology will be a key tool for accommodating the millions of inputs, contracts, and choices that are coming online from renewables and connected home and industrial devices.

Renewable energy is entering the grid at an unprecedented rate at the same time that electric cars and other machines on the IoT are making their power demands known. Blockchain has the potential to securely manage an infinite number of transactions, and it will be a game changer for utilities and others as the grid modernizes with renewable inputs. After all, managing the energy of thousands of homes' solar or wind generators is a different story from managing the output of a large coal-fired plant. Blockchain will be able to handle it.

The grid is a web of connections that these days must handle power flowing in both directions, and also track payments to and from a legion of independent producers. Lately, this complex system has found itself challenged by electric vehicles, home solar panels, smart refrigerators, and, of course, smartphones. All these devices now communicate their power needs, and could become powerful energy assets, using blockchain.

As I've mentioned, blockchain is often criticized for the environmentally destructive amounts of electricity required to settle some types of transactions, especially those involving bitcoin. However, the ledger also offers some potentially game-changing solutions to other problems that contribute to climate change. Many knowledgeable people see carbon-trading markets that put a price on carbon pollution as a great way to reduce the amount of emissions, while earning money to offset the problems we already have. For instance, IBM and Energy Blockchain Lab have developed a carbon asset trading platform, using Hyperledger, that is expected to simplify and speed up transactions in the Chinese market and make the market more transparent.

Other interesting applications of blockchain for climate change, and other environmental purposes, include:

- More accurate and transparent greenhouse gas emission tracking, worldwide.
- Easier access to funds to fight climate change and develop new technologies.
- Arming wild areas with sensors that can help scientists remotely monitor the health of the planet.
- Trading platforms for clean energy.
- Ease in onboarding renewable energy into the grid, from a variety of sources.
- The possibility of transparently monitoring the progress of pledges nations made under the Paris Climate Agreement.

- Certifying ownership of a forest or water rights, to prevent unwanted incursion or development.

- Tracking endangered animal parts, minerals, and other commodities so they aren't sold around the world.

- Tracking the use of funds that are donated for environmental purposes, to show whether they are properly used.

The UN, the World Bank, and other large institutions are now working on projects to harness the positive effects of blockchain on the environment. At the same time, many technological efforts are under way to reduce power usage by blockchains.

Next

... the wicked lie, that the past is always tense,
the future perfect.

—*Zadie Smith*

Blockchain utopians predict a bright future for the planet, as world-changing uses of the technology transform society. Clearly, blockchain has the potential to influence and even transform a long list of industries, concepts, and systems. Here are a few examples of how this might happen:

- The Internet of Things (IoT) refers, as described, to our increasing practice of connecting objects like televisions, cars, toothbrushes, and industrial machines to the Internet. Blockchain tech, particularly smart contracts, will make it easier for these things to handle transactions, make requests, and fulfill the needs of other machines and humans. For instance, your refrigerator might be able to order milk for you, and a tire production line might be able to calibrate its speed and output automatically according to signals it gets from new car sales.

- Financial transactions will be faster without the need for repeated assurances from mediators and gatekeepers.

- Artists might more easily receive royalties on the resale of their artworks, as blockchain is used to certify provenance and track sales. Collectors could easily form consortia to purchase expensive works of art, with all the details handled by smart contracts.

- Everything from energy use to materials waste will become more efficient, greatly reducing carbon emissions that cause climate change. Communities that steward resources well could be rewarded using blockchain tech. Smart cities could encourage many behaviors, such as sharing transportation and reducing water use.

- Charities will benefit from increased transparency and lower transaction costs.

- Health care records and resulting privacy issues are ripe for blockchain innovation, as is the entire health insurance landscape.

- Already entrepreneurs are using blockchain tech to facilitate fractional ownership of properties, to give more people access to the tremendous wealth in real estate.

- In much of the world, land titles are difficult to obtain and maintain. Blockchain tech can solve this.

- The retail environment is due for a change; imagine a blockchain that held records for every item of fashion for sale on earth—letting you pick and choose what you wanted. Imagine such a blockchain that listed only sustainable clothing, so that you'd never have to wonder about the ethical challenges of what you are buying.

- Cryptocurrencies could change the very concept of value, helping our society adjust to jobs lost due to innovation and automation. Universal basic income, which many people now categorize as a "handout," would instead be a reward for behaviors—helping in the community, taking care of your family, composting—that benefit society. Blockchain could reward these behaviors with different types of coins.

Still, all good things come to an end. Blockchain technology will most likely be superseded in the future by something that works even better. In a way, blockchain may be a self-negating force. As the infrastructure of a new social, political, and financial order, where hierarchies have less power and hive collaborations have more, blockchain will allow new technologies to rise to the surface that may replace it. Blockchain is only the beginning of our transition into distributed systems that promise to transform how we do business, communicate, and govern. The creative capacity unleashed by distributed systems like blockchain will speed the change.

Blockchain week, 2018. Headlines from the *New York Post*, a tabloid that goes to great lengths to get attention:

Bitcoin Has Become Boring
(bemoaning bitcoin's recent price stability)

Nouriel Roubini Says Bitcoin Is "Bulls—"
(who knows, the New York University economist might have a point)

Cryptocurrency Could Bring Chaos to U.S. Economy
(top Federal Reserve dude issues a warning)

Inside the Crypto Bro Fest That Took Over New York City
This one's illustrated with a photograph of several Lamborghinis parked in front of a blockchain conference in NYC's Blockchain Week, which features conferences, debates, lots of parties, and many Lambos—the crass automotive symbol of cryptomillionaires, of which there seem to be many.

Aside from the Ethereal Summit out in Queens, which seems pretty serious about fixing the world, the week evokes strong memories of the end of the Internet bubble at the turn of the century. I remember one startup back then that delivered anything you wanted to your door, for free. I'd call them for lunch every day, and they'd drop it off. No tipping. Sick. They were trying to gain a larger share of the market. But market share of what? Freeloaders?

For a crazed few months back then, young, newly wealthy Internet bubblers were throwing crazy parties, night after night. At Blockchain Week it's much the same thing, sometimes including some of the same musicians who performed back in 1999. One of the latest parties, at a nightclub on the West Side, featured a performance by Snoop Dogg. I mean, pot is legal, rap has moved on, he might as well be Desi Arnaz. Parties everywhere. Nude dancers. Dom. Coin talk in bathrooms, as people sniff coke. It's hard not to feel like a bubble is near the bursting point.

But that bubble is for speculation in cryptocurrency, not for solid thinking about how blockchain innovations can affect industry and society. Blockchain's not in a bubble at all.

Deepak says, "Blockchain is a human construct, becoming an every-day reality."

Well, duh, I think, before deciding to be more generous with my thoughts. I read a Deepak Chopra book a long time ago, and I thought he was pretty wise. He's standing at the front of a low stage, holding a microphone. He looks as if he'd be fun at a party. His sneakers are dope, black with bright red socks.

All of the blockchain bros wear dope sneakers, especially the ones who are speaking on stages at this Ethereal Summit, at the outer edge of New York City, in Maspeth, Queens. It's not so far from the head-quarters of ConsenSys, the blockchain studio sponsoring the summit, which was founded by Joe Lubin, who cofounded the Ethereum chain and now, apparently, is a friend of Deepak.

I'm in a wooden chair. Between me and Deepak are young people lounging on plush floor pillows, women mostly, wearing leggings and boots, a few sneakers. Behind and around and on top of us, the roar of the thousands of people attending this conference who are on the other side of the thin walls, networking and listening to speakers. I see beau-tiful sweaters and glass phones with no cases, rabbit fur Borsalinos. A lot of coin wealth in this room. I'm in the conference's Zen Zone and I feel coin envy. Why the hell didn't I buy when . . . I kick myself for not buying bitcoin five years ago—if I had I'd be a rich man now—and return my gaze to Deepak, with his dark Nehru jacket, the collar and pocket piping bloodred to match his socks.

He leads us in a meditation, and despite the disembodied roar behind me, I fall into it. Minutes later, the rain is beginning to fall outside and Deepak opens the meditation to questions and comments.

A young man sitting cross-legged, back rigid straight, offers a comment: "I have been studying yoga for about the same amount of time that I've been investing in crypto." His voice is modulated, deep, perfectly attuned with the harmony of the universe. "I have begun to see parallels between the system of the chakras and blockchain and the foundation of the new distributed Internet."

Deepak nods, a little smile on his face. I'm wondering, is Deepak pondering this revelation? Is Deepak thinking, that chakra stuff is my bailiwick, I had that thought yesterday? Or is Deepak calculating how much his bitcoin will be worth when it goes to thirty grand?

"Do you see a connection, Deepak?" the young man says, hands in prayer mode and bowing with humility.

"It could be, it could be," the master replies.

Outside, thunder claps. I gather my wits and check my map. I need to get to the Jefferson Avenue station. It's right there, linked to all the other station nodes in the city by different-colored connectors. I think, oh, yes, of course, this is how it works—distributed system. Or is it?

A gaggle of youths blocks the sidewalk as I head to the station.
Bomber jackets, nose rings, pink and green and yellow hair, the mashup
of cultures you so often see in groups of NYC kids—our future. One
young woman wears a T-shirt that reads "Satoshi is Female."

There's so much more to this stuff than cash. So much room for
these kids to create new worlds. Blockchain is a political issue. A social
justice issue. An economic issue. A business issue. A sustainability
issue. It's a transformative issue. As all the blockchains start to communicate, and all the nodes are linked together, the new Internet is
born, free of old constraints and ownership.

And I'm hoping we can ride it right into the new enlightenment.

Acknowledgments

I thank the giant distributed system of New York City that somehow allows disparate thinkers to connect in serendipitous ways—I've learned a lot about blockchain this way. To John Capouya, great friend and longtime editor, who gave this project 100 percent and also improved it by that much. My gratitude to the Bard MBA in Sustainability program and Hunter Lovins's class, where I first used the word *blockchain* in public. And to Melanie Swan, whom I've never met, but whose visionary words inspired me. Thank you blockchain-and-crypto Twitter for all the informative and at times puzzling info via blue light late at night. In that blue light I often encountered Ben Schiller, editor of Breakermag.com, which has been generous enough to publish my stories about blockchain. To Rosemary Ryan, of co:collective, for seeing, astonishingly early, that blockchain would eventually pervade business culture and telling me to plant my stake. Reagan Richmond, my partner in various distributed ventures, has been a steady, clear-headed adviser and motivator. Thanks also to Josh Quittner, of Decryptmedia.com, for getting me into the Ethereal Summit and for publishing lots of great info. I'm forever grateful to my brothers, Jonathan and Peter, for listening without audible yawns to my proud complaints about being too busy.

Much appreciation and respect to my gifted, patient, and always positive editor, Daniel Loedel, who saw the book exactly as I envisioned it and brought it to life. Also to Dani Spencer, my hardworking publicist, and Katie Rizzo, my marvelous production editor. And to Liz Perl, the CMO of Simon & Schuster, who has a good eye for good tech and a great way of bringing people together—thank you, Garth Holzinger, for introducing us and leading good conversations. I have so much appreciation for my wonderful agent, Susanna Lea, and her associates Noa Rosen, Devon Halliday, Lauren Wendelken, Kerry Glencourse, and Mark Kessler. Their hard work and appreciation made this book possible. Finally, thank you to Aspen, Violet, and Bolivia for unflagging encouragement and cups of tea.

Further Reading

Breakermag.com is a terrific source of colorful information about the culture, glory, and despair of the blockchain world.

Goldman Sachs, the investment bank, has a great graphic presentation on blockchain tech. I marvel at its simplicity and grace. Google "Blockchain: The New Technology of Trust" to see for yourself.

Blockchain Revolution: How the Technology Behind Bitcoin Is Changing Money, Business, and the World by Don Tapscott and Alex Tapscott was the first intelligent, readable book I read on the subject. It goes broad and deep.

Ethereum cofounder Vitalik Buterin published a fascinating, if occasionally opaque, article on *Medium* titled "The Meaning of Decentralization." It's worth reading for two reasons: Buterin is a lot smarter than I am, and decentralization and distributed networks are the big ideas within blockchain tech, and most people in the space don't think enough about them.

Decryptmedia.com offers straightforward, easy-to-read news about crypto and blockchain, including an educational arm called Lite-Paper: Blockchain & Cryptocurrencies Made Effortless.

Blockchain.com/explorer lets you search actual blockchains.

Wired published "The Blockchain: A Love Story/The Blockchain: A Horror Story, Inside the Crypto World's Biggest Scandal" by Gideon Lewis-Kraus in June 2018. It will teach you a lot.

Philosopher and futurist Melanie Swan's website, www.melanieswan .com, has a plethora of interesting links to speeches, slide decks, papers, and books about the potential of blockchain.